APOLLONIAN EXHIBITION
ICONOGRAPHY OF THE APOLLONIAN SYSTEM

Conceptualizing the Unseen

Written and Illustrated by LANCE BURRIS

Cover design and paintings by the author. Text formatting and photography by L2 Napa. Epoxy-resin finishing of original paintings by Andrea Cazares.

Printed in the United States of America. ISBN **9798343780079**

Copyright issued in 2024 by Lance Burris as the author of this book to whom all rights are reserved. No part may be used or reproduced in any manner without the written permission of the author except in the case of brief quotations embodied in critical articles or reviews. For additional information write to Lance Burris by email at lburris101@gmail.com.

"Science is no more than the investigation of a miracle we can never explain, and art is an interpretation of that miracle." Ray Bradbury

OPENING REMARKS

This book serves as an interpretive guide to the Apollonian Exhibition. The Apollonian Exhibition introduces a new art form based upon the principles of four-dimensional geometry and the mathematics of the infinite. It consists of a "Concept Box" and a "Folio." The former describes my metaphysical philosophy, which I call the Apollonian System. The latter consists of my twenty-two oil paintings of the Apollonian System's conceptual icons, which are archetypes designed to instill an intuitive sense of the Apollonian Multiverse as an alternative to the current scientific model of reality.[1]

It is impossible to visualize objects in four-dimensional space because we lack a perceptual basis for doing so. However, the Apollonian System's conceptual icons enable one to conceptualize the Apollonian Universe as an integral part of a greater Multiverse. It does so from the perspective of our consciousness which looks down from a fourth-dimensional height upon a section of the three-dimensional physical world.

The famous twentieth century mathematician Alexander Grothendieck once observed: *"A perspective is by nature limited. It offers us one single vision of a landscape. Only when complementary views of the same reality combine are we capable of achieving fuller access to the knowledge of things. The more complex the object we are attempting to apprehend, the more important it is to have different sets of eyes, so that these rays of light converge, and we can see the One through the many. That is the nature of true vision. It brings together already known points of view and shows others hitherto unknown, allowing us to understand that all are, in actuality, part of the same thing."* The Apollonian Exhibition provides that "true vision" and "different sets of eyes." It enables us to conceptualize rather than "see," as Grothendieck said, different aspects of the Apollonian Universe and Multiverse and, by doing so, develop a feeling for the whole. The purpose of the Apollonian Exhibition is to instill an intuitive sense of the perceptual consequences of imposing our single point of view and the three-dimensional receptivity of our bodily

---

[1] The Apollonian Multiverse consists of an infinite number of consciousness-centered Apollonian Universes. These universes are conceptualized in the collective as an eternal, four-dimensional Möbius Strip in which the strip represents the physical world, which is finite in being, and its single surface represents the perceptual world, which is infinite in nonbeing. The term "being" implies a fourth-dimensional conscious presence.

senses upon four-dimensional events. Mathematical manipulation alone cannot overcome our inability to perceive directly the physical world, which Kant called "the thing in itself." When String Theorists speak of eleven dimensions, they have no basis in perception for visualizing objects in this higher-dimensional space. The Apollonian System's conceptual icons provide glimpses of the greater reality which underlies our limited perception of it in space/time past.

While there could be an infinite number of dimensions, only four are necessary to explain our experience of the phenomenological world. We know we are four-dimensional beings because we must be at least one dimension removed from that which we perceive in three-dimensions. The Apollonian Exhibition's conceptual icons provide mental manipulatives which allow us to think in the four-dimensional space in which past, present, and future coexist. While these archetypes are two-dimensions removed from what they seek to portray, they embody the principles of four-dimensional geometry and the mathematics of the infinite which allow us to mentally navigate four-dimensional space. This expanded frame of reference makes it clear that perception is not reality any more than visual perspective is Euclidian. Therefore, we should interpret our perception based upon the single point of view and fourth-dimensional perspective of our consciousness. The Apollonian Exhibition uses the indirection of art to instill an intuitive sense of both.

CONTENTS

OPENING REMARKS iii

CONTENTS

CONCEPT BOX
    General Description 5
    Front Triptych 7
    Rear Triptych 9

FOLIO
    General Description 13
    Master Conceptual Icons:
        MCI-1: Cosmology 17
        MCI-2: Ontology 19
        MCI-3: "Bohr Atom" of the Conscious Universe 21
        MCI-4: Epistemology 23
    Allegorical Conceptual Icons:
        ACI-1: Point of View 27
        ACI-2: Mechanics of Perception 29
        ACI-3: Creation of Knowledge 31
        ACI-4: Evolution of Consciousness 33
        ACI-5: Particulate Matter 35
        ACI-6: Invisible Offset 37
        ACI-7: Thou Art That 39
    Working Conceptual Icons:
        WCI-1: Apollonian Coordinate System, Event in Being 43
        WCI-2: Apollonian Coordinate System, Event in Transit 44
        WCI-3: Apollonian Coordinate System, Perceptual Event 45
        WCI-4: Wave and Particulate Nature of Light 47
        WCI-5: Expanding Sphere of Consciousness 49
        WCI-6: Contracting Sphere of Perception 51
        WCI-7: Ring Diagram Analysis 53

  WCI-8: Venn Diagram Analysis 55
  WCI-9: Geometry of Probability. 57
  WCI-10: Fibonacci Spiral 59
  WCI-11: Scale of Being 61

CLOSING REMARKS. 63

APPENDICES

 A. Rear Triptych Text 67
 B. Allegories
  No. 1—The Old Man in the Barn 77
  No. 2—Vision at Cobá 79
  No. 3—Nemo's Corollary. 83
  No. 4—Road Race. 89
  No. 5—The Medium is the Message. 91
  No. 6—Dark Studio. 93
   No. 7—Infinity Mirror 95
 C. Exhibition Plan. 99
 D. About the Author/Artist 101

CONCEPT BOX

General Description

The Concept Box has two breakfronts forming back-to-back triptychs which, when opened, display information about the Apollonian System and Exhibition. The box's purpose is to preserve more than forty years of my independent philosophical thought in a single physical artifact.

Front Triptych

Contents:

- The left-hand pocket contains a copy of my book titled: *Apollonian Coordinate System, the Mechanics of Perception from a Fourth-Dimensional Perspective. Toward a Deductive Metaphysical Science of Conscious Being.* The book summarizes the findings of earlier unpublished manuscripts documenting the Apollonian System's purpose, formal logic, mathematics, and visual vocabulary.

- The central panel displays my oil painting of the ruins of the Temple of Apollo at Delphi in Greece (i.e., the *Apollonion*) upon which the admonition "know thyself" is said to have been inscribed. I use the attribution "Apollonian" because the System's primary purpose is self-knowledge.

- The right-hand pocket contains a copy of this book which is titled: *Apollonian Exhibition, Iconography of the Apollonian System. Conceptualizing the Unseen*. The book describes the Apollonian Exhibition's Concept Box and Folio of conceptual icons, which are intended to instill an intuitive sense of the mathematically based metaphysics of the Apollonian System.

Rear Triptych

- Contents:[2]

- The left-hand panel displays my pen and ink drawing of the Oracle of Delphi (aka, the Pythia) and my poem titled the *Sea of Light*. The poem was written in my youth and anticipates my future development of the Apollonian System's theories of being and knowledge, color-code, four-dimensional coordinate system, and new paradigm for time, space, and light.

- The central panel describes the Apollonian System and Exhibition.

- The right-hand panel lists the titles of the Apollonian Exhibition's paintings and related allegories and explains the Apollonian System's color-code.

---

[2]See Appendix A for full text.

FOLIO

General Description

The Folio consists of twenty-two oil paintings of the Apollonian Exhibition's master, allegorical, and working conceptual icons. Each is painted on a 20-inch square, wooden panel and color-coded in accordance with the Young-Helmholtz Theory of Human Color Vision.[3]

---

[3] The theory states that an admixture of red, green, and blue light creates white light. In the paintings, red represents space/time past, blue represents space/time future, and green represents the present being which is the domain of the physical world. White and black represent conscious being and non-existence, respectively.

MASTER CONCEPTUAL ICONS

Description

The master conceptual icons display the geometric structures of the Apollonian Universe and greater Apollonian Multiverse in section. The Apollonian Universe (also called the microcosm) is centered on individuated consciousness. The Apollonian Multiverse (also called the macrocosm) is centered on universal consciousness. Universal consciousness is tangent to the plane of being of the physical world where it manifests as individuated consciousness (also called the self). These geometric structures answer the ancient philosophical question of the "One and the many" which accounts for the origin and nature of individuation.

MCI-1: COSMOLOGY
Topology of the Apollonian Multiverse

This conceptual icon displays a section of the Apollonian Multiverse which is in the form of a four-dimensional Möbius Strip consisting of universal consciousness in an eternal state of being, the finite physical world in a state of present being, and the infinite space/time past of the perceptual world and infinite space/time future of the world of probabilities in states of nonbeing.[4] The icon is also known as the *Burris-Van Manen Cosmos* because it is a composite of the Apollonian System's mathematically generated horns of existence and Van Manen's mystical vision of a four-dimensional shape.[5]

[4] A Möbius Strip is created by giving a strip of paper a half-twist and joining the ends. The resulting figure has a single surface which feeds back into itself like a snake eating its tail. The surface of the strip is a model of the infinite space/time of the perceptual world, while the strip itself is a model of the finite physical world.

[5] I first encountered the name of the Orientalist Johan van Manen in P. D. Ouspensky's book titled: *Tertium Organum. The Third Canon of Thought.* The book contained a drawing of Van Manen's vision of a four-dimensional shape as it appeared to him in that vision. I discovered the drawing resembles the Apollonian System's horns of existence which I had derived mathematically. In fact, it is identical when the ends of the horns are connected to form a four-dimensional ring diagram. The Apollonian System's horns of existence are generated by rotating the Curve of Aretê through a third dimension. The Curve of Aretê is created by varying the fourth-dimensional heights of the Apollonian System's cones of perception and probability.

Cosmology consists of the following:

- A white circular area representing universal consciousness which is the source of awareness, the will, and being. While universal consciousness is represented mathematically by a point, it is shown as a circular area for purposes of illustration.
- A green ring encircling the white circular area which represents the three-dimensional physical world, each-and-every point in which is grounded in being by universal consciousness.
- A red horn of existence representing the space/time past of the perceptual world, to which the abstraction of time has been added at each point in three-dimensional space to create unique events. Perceptual events in space/time past can appear to move; however, this motion is illusory because it is simultaneously offset by an equal and opposite displacement in space/time future. This invisible offset is the product of hyper-spherical expansion which results in no net motion in four-dimensional space we experience as the "now" of present being.
- A blue horn of existence representing the space/time future of the world of probabilities, which is the counterpart of space/time past and constitutes the invisible domain of dark energy and matter.[6]
- Nonexistence is shown in black.

---

[6] The fields of space/time past and future are in states of nonbeing relative to the observer's consciousness which is embedded in the plane of being of the physical world. They are created by the will-driven process of thought resulting in prime motion which produces hyper-spherical expansion at a speed equal to that of light. Our mind-bodies give expression to this expansion as physical activity and perceived as the propagation of light and experienced as the passage of time. Displacement of an unseen quantum, which a unit of consciousness, in the fourth dimension is the source of what physicists call energy. From our consciousness' single perspective, objects in the three-dimensional subspace appear to rotate while undergoing prime motion. These mechanics generate the infinite space/time of the perceptual world which forms the closed loop popularly referred to as the space/time continuum. The three-dimensional aspect of this space/time field collapses to an infinity point every rotation of pi-radians of arc. In effect, our perceptual limitations digitize the space/time past of the perceptual world. This process accounts for the dual wave and particulate nature of light in which particles periodically emerge which gives the appearance of discontinuity. At each infinity point, the vector of time is reversed relative to the plane of being in which our consciousness is embedded. This reversal makes the Apollonian Universe and greater Multiverse a self-contained and eternally recycling unified field (i.e., a singularity) in which all is in all, each is in all, and all is in each.

MCI-2: ONTOLOGY
Origin and Nature of Individuation

This conceptual icon provides a geometric model of Apollonian Universe and greater Multiverse. It demonstrates how the One of universal consciousness (Ψ) becomes the many of individuated consciousness (ψ), also known as the self, the latter being a fourth-dimensional unit of consciousness which physicists call a quantum. The quantum contains the informational "DNA" of the physical world and, as such, is the Platonic ideal which manifests as form in the physical world. The quantum manifests in the three-dimensional subspace as a photon which is the fundamental building block of the material world as perceived. Prime motion translates the quantum's displacement in the fourth dimension into hyper-spherical expansion, differential rotation, and particulate spin in the three-dimensional aspect of the space/time field of the perceptual world.

Ontology consists of the following:

- A white circular area representing universal consciousness, which is the eternal source of awareness, the will, and being. While universal consciousness is described mathematically by a point, it is shown as a circular area for purposes of illustration.

- The smaller white dots encircling the white, circular area, represent individuated consciousnesses. Each is embedded in the green plane of being of the three-dimensional physical world which is shown in linear section. This conceptual icon accounts for the origin and nature of individuation by demonstrating how the One becomes many. Each self can be said to be in the world but not of it, which means consciousness is not an epiphenomenon of the physical world as science believes because of the scientific method's underlying materialist philosophy. Rather, it is a manifestation of universal consciousness which is the source of the physical world's being.

- Four green lines representing the physical world which is shown as a cube in linear section. A cube is used in the interest of simplicity to represent an infinitely sided polyhedron, the sides of which approach a spherical surface as a limit.

- Red and blue hemi-hyperspheres representing the space/time past of the perceptual world and space/time future of the world of probabilities, respectively. Past and future are measured relative to the plane of being of the physical world into which individuated consciousness is embedded and divides hyper-spherical space/time. Space/time is generated by an inertial system consisting of this plane of being and embedded self which undergo uniform and continuous prime motion in the fourth dimension at a speed equal to that of light. The plane of being is merely a construct of the self in which points of tangency with universal consciousness are arrayed in two dimensions for ease of reference. These points of tangency become one and the same when rotated through pi radians of arc in the third and fourth dimensions. Each of the points represents the face of universal consciousness. This is consistent with the Hindu declaration: "thou art that." Each point in the three-dimensional physical world has a conscious fourth dimension which is the source of its being.

- Cones of perception are shown in triangular section colored a darker shade of red, and cones of probability are shown in triangular section colored a darker shade of blue. Red and blue lateral elements of the cones of perception and probability constitute perpendicular lines of space/time past and future, respectively. These lines of space/time form a network along which light travels and connects the Apollonian Universes within the Multiverse.

- Nonexistence is shown in black.

MCI-3: "BOHR ATOM" OF THE CONSCIOUS UNIVERSE
Four-Dimensional Frame of Reference

The Danish physicist Niels Bohr created a conceptual model of the hydrogen atom consisting of a single electron orbiting a single proton. This model was instrumental in the development of the science of Quantum Mechanics and related Standard Model of Particle Physics. The model deepened our understanding of the subatomic world and led to the realization that the experimenter's consciousness somehow influences the perceived experimental results. However, the materialist philosophy of the scientific method prohibits Quantum Mechanics from formally acknowledging the reality of that conscious presence. The *"Bohr Atom" of the Conscious Universe* addresses this oversight by embedding the observer's consciousness in the plane of being of the physical world at a point fixed in absolute four-dimensional space. By doing so, it establishes a central point of reference in the hyper-spherical Apollonian Universe as an integral part of a multi-centered Multiverse governed by the principles of four-dimensional geometry and the mathematics of the infinite. The Apollonian System uses a section of the "Bohr atom" of the conscious universe to establish the spatial frame of reference for the four-dimensional Apollonian

Coordinate System, the origin of which is located at a fixed point in four-dimensional space and represents consciousness in being.

The *"Bohr Atom" of the Conscious Universe* consists of the following:

- Red and blue diagonal lines representing lines of space/time past and future, respectively. These perpendicular lines are lateral elements of the cones of perception and probability which are shown in section in triangular outline.

- Red and blue horizonal lines representing parallel lines of existence past and future relative to the plane of being of the physical world which is shown in section as a vertical green line. A line of existence is generated by a quantum undergoing prime motion in the fourth dimension at a speed equal to that of light. Again, a quantum is defined as a unit of consciousness which is the source of being in the physical world.

- The consciousness of the observer is fixed at the origin of the Apollonian Coordinate System where it establishes a central frame of reference in the four-dimensional Apollonian Universe which is an integral part of the greater Multiverse. The central white dot represents the self as a manifestation of universal consciousness embedded in a green linear section of the physical world. The white dot is located at the intersection of the red, green, and blue lines in accordance with the Young-Helmholtz Theory of Human Color Vision.[7]

- Red and blue vertical lines represent radii of the circular planes of perception and probability constituting the bases of the cones of perception and probability, respectively.

- Red and blue concentric circles represent sections of iso-space/time past and future and sections of spheres of perception which collapse upon the observing self as the physical world and self undergo prime motion in the fourth dimension at a speed equal to that of light.

- Black represents nonexistence.

---

[7] The Young-Helmholtz Theory states that red, green, and blue are the fundamental components of human color vision. An admixture of equal intensities light of these colors is perceived by the human eye as white light. Red and blue are also used as a mnemonic device for the receding past and approaching future, respectively.

MCI-4: EPISTEMOLOGY
Creation of Knowledge

This conceptual icon demonstrates how our *a posteriori* knowledge of the external world is based upon our *a priori* assumptions about which determine the mechanics of perception in four-dimensional space. The scientific method is based in perception which makes our knowledge of the physical world derivative and subject to interpretation based upon unprovable metaphysical assumptions. The only thing we truly know is our own consciousness because it is the only "thing" we directly experience in a state of being. If we do not understand the mechanics of perception from our limited perspective as four-dimensional conscious beings, we simply don't know what we are looking at. This lack of understanding leads to the misinterpretation of our perceptual experience, and we mistake perception for reality when they are not the same.

Epistemology consists of the following:
- A vertical green line representing a section of the plane of being which, when rotated through a third-degree of freedom, generates the three-dimensional space of the physical world.

- A white dot representing the self as individuated consciousness. This self is in the fourth dimension which is located perpendicular to the plane of being into which the self is embedded.

- Perpendicular, diagonal red and blue lines of space/time past and future intersect the self. Light propagates along these lines which constitute lateral elements of the cones of perception and probability, respectively.

- A green star, representing an event in being, is located at a fixed point in the absolute four-dimensional space of the Apollonian Universe.

- A blue star, representing the physical source of the event which is located in space/time future where it cannot be seen.

- A red star representing the perceptual event occurs when the self and the event are in a shared state of being at a fixed point in four-dimensional space. This shared state is represented by a green star with a white dot at its center. The mind-body registers this shared state of being on the consciousness of the observer as visible light which enables the self to see the event in the space/time past of the perceptual world.

- A black star outlined in green represents a virtual event which appears to be contemporaneous with its observation but is not.

- Red and blue sections of the horns of existence past and future, respectively, which are generated by rotating the Curve of Aretê through two pi radians of arc. Under this curve, the bases of the cones asymptotically approach the area of the plane of being as a limit as their altitudes approach zero. In this manner, the self accumulates *a posteriori* knowledge of the world as measured on the plane of being. The Curve of Aretê is generated by varying the heights of the cones of perception and probability as the self exercises its will through the agency of the mind-body. The radii of the bases of the cones are measures of the self's level of consciousness which approaches truth in being as a limit.

- Black represents non-existence.

ALLEGORICAL CONCEPTUAL ICONS

Description

Allegorical conceptual icons are based upon fictional stories which serve as thought experiments and instill a feeling for the geometric principles and dynamics governing the Apollonian Universe and Multiverse. When attempting to comprehend the unfamiliar, insights can be obtained by drawing a comparison with the familiar.

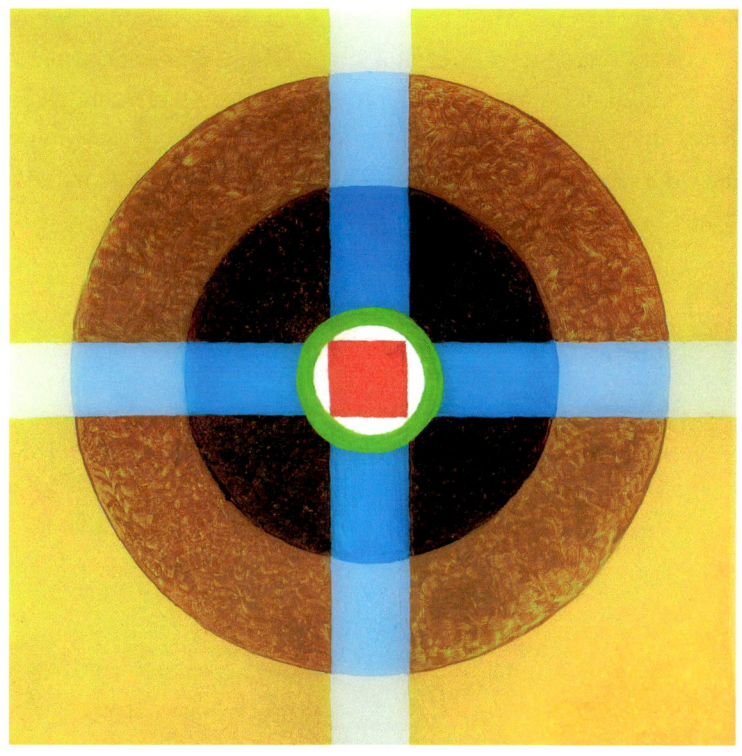

ACI-1: POINT OF VIEW
Allegory No. 1—*The Old Man in the Barn*

This conceptual icon is based upon the allegory titled *The Old Man in the Barn* (see Appendix B-1 for text) in which an older man suffering from Alzheimer's disease wanders away from an institution and enters an abandoned circular barn in the nearby countryside. Attracted by the light from the nearest window, the old man feels his way along the barn's interior wall, moving from window to window as the biochemistry of his memory breaks down. During his journey into forgetfulness, he views the world in a succession of soon forgotten images, each of which becomes a separate reality. In the icon, the barn wall and floor are colored lighter and darker shades of brown, respectively; the dry grass in the countryside is colored yellow. The windows are colored blue indicating their location in space/time future. At the center of the icon, the old man's mind-body is shown as a circular green area, indicating a state of present being. A beam of blue light streams toward him as an event in transit. A red square, centered in the green circle, represents the image from a single window as it registers on the old man's consciousness as an event in space/time past. In this manner, a combination of barn geometry, bodily motion, and failing memory creates separate realities in a single image as if experienced by different observers, each of which has a different point of view. Yet, there is only one consciousness in the barn. This suggests our three-dimensional mind-bodies share a common consciousness in the fourth dimension. Since this connection cannot be seen, we experience ourselves as separate individuals. This suggests a shared universal consciousness looks down upon the physical world through our eyes.

ACI-2: MECHANICS OF PERCEPTION
Allegory No. 2—*Vision at Cobá*

This conceptual icon is based upon the allegory titled *Vision at Cobá* (see Appendix B-2 for text). The allegory was inspired by my visit to the great pyramid at Cobá on the Yucatan Peninsula, during which I imagined myself being transported back to a day when Maya civilization was at its peak. As a result of that, I conceived of the idea that the pyramid was a model of perception which is activated by ritual. The icon consists of a red pyramid which encases the cone of perception in space/time past. The red pyramid is surmounted by a blue one which encases the cone of probability in space/time future. A central staircase leads to the top of the red pyramid, while another descends the inverted blue pyramid in the sky which is invisible to the human eye. A standing sine wave undulates up the red pyramid's staircase and down the blue pyramid's staircase which are aligned with lines of space/time past and future, respectively, defined by the lines of tangency with the underlying cones. The sine wave traces the path of a procession of Maya nobles and warriors as they climb to the top of the pyramid where the physical act of human sacrifice is performed as the price of admission to the inverted pyramid in the sky, the staircase of which must be descended for Maya civilization to live on. A horizontal green line represents the plane of being of the physical world, and the white dot represents the consciousness of the sacrificial victim as a representative of the Maya race.

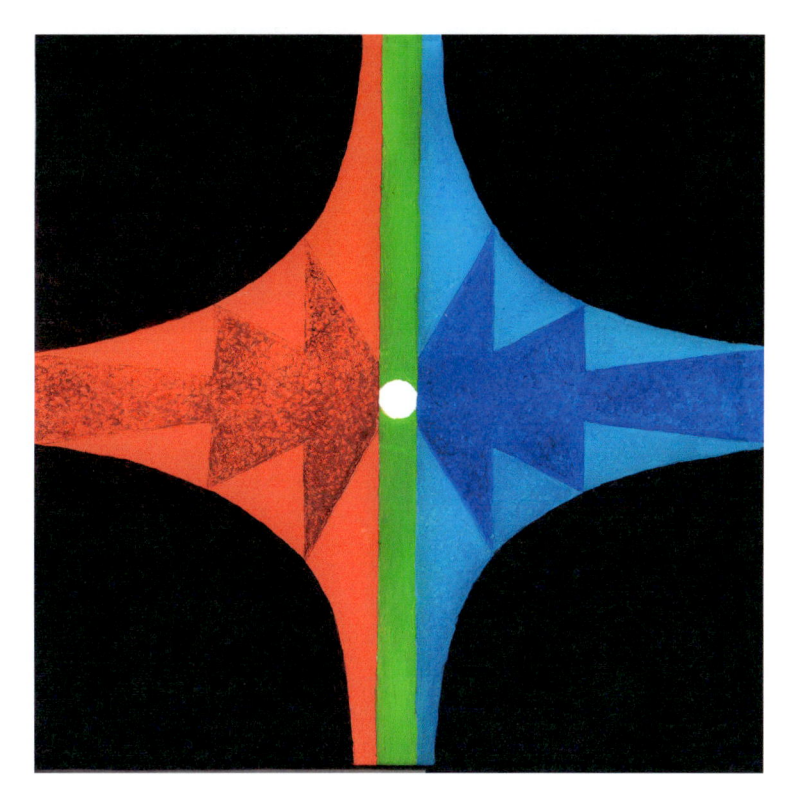

ACI-3: CREATION OF KNOWLEDGE
Allegory No. 3—*Nemo's Corollary.*

This conceptual icon is based upon the allegory titled *Nemo's Corollary* (see Appendix B-3 for text). The allegory describes a conversation between a high school student and his geometry teacher. During the conversation, the teacher tells the student that uncertainty underlies mathematics, science, and religion because all are based upon unprovable metaphysical assumptions. In the icon, the cones of perception and probability have fixed volumes. In the case of Man, each is arbitrarily set equal to one, since Man is currently the only known creator of culture and civilization. Therefore, Man becomes the standard against which all other conscious beings are measured. The juxtaposed cones are shown in section and colored red and blue, representing space/time past and future, respectively. The individuated consciousness of the self is shown as a white dot which is embedded in the green plane of being of the physical world. By force of will, the self is able to achieve higher levels of consciousness by the process of knowledge creation. This dialectical process shortens the heights of the cones such that the area of each base approaches the area of the plane of being as a limit as perception approached truth in being. This process of knowledge creation generates a Curve of Aretê by which the level of consciousness of the self asymptotically approaches infinity (i.e., all knowingness) as a limit.

ACI-4: EVOLUTION OF CONSCIOUSNESS
Allegory No. 4—*Road Race*

This conceptual icon demonstrates how accumulated knowledge raises the consciousness level of the individual or species. Acting on this knowledge is not without risk, however. The related allegory, titled *Road Race,* describes an automobile race involving three racing teams sponsored by three different automobile manufactures. While the cars have different drivers, they are mechanically identical by team but differ from team-to-team. However, the cars have one thing in common. In each case, the windshield and side windows are painted black. This means the drivers are only able to view of the road as it existed in the past as shown in the rearview mirror. Therefore, each driver is forced to project a mental image of the road ahead based upon the perceptual information obtained from the rearview mirror and the assumed continuity of space/time. This situation is a metaphor for the uncertainty inherent in our four-dimensional nature as conscious beings. In the icon, the image in the rearview mirror is shown as a red rectangle in space/time past. The racing car's mirror and window frames, steering wheel, and dashboard are colored green indicating their state of present being in the physical world. A blue triangle represents the driver's assumed perspective of the road ahead based upon his perception of it in the past. This means he drives blindly into the future, with uncertainty being a fundamental attribute of the human condition.

ACI-5: PARTICULATE MATTER
Allegory No. 5—*The Medium is the Message*.

This conceptual icon is based upon an allegory titled *The Medium is the Message* which echoes Marshall McLuhan's famous maxim. The allegory describes the *quipu* which was a unique communication system used by the Incas and consisted of a bundle of knotted strings. The size, number, and spacing of the knots stored information digitally. In the allegory, the quipu serves as a metaphor for the structure of particulate matter which is the product of the limited receptivity of our bodily senses. In other words, matter's particulate nature is illusory and not a physical reality in four-dimensional space. The horizontal blue line in the icon represents a segment of a continuous line of space/time future. The undulations represent a sectional view of the segment's rotation in three dimensions which we perceive as separate physical particles because we cannot see their fourth-dimensional aspect. The green color indicates the particle is in a state of present being. The winding density of the line of space/time determines the particle's mass and equivalent energy measured in units equal in magnitude to Planck's Constant. As four-dimensional conscious beings, we are unable to perceive displacement in the fourth dimension because our consciousness is always aligned with that dimension—that is to say, it is not one dimension removed. This alters the space/time ratio which determine the speed of light which distorts the perceptual field. We experience this distortion perceptually as the compression of space and physically as a force such as visual perspective and force of gravity, respectively. We simply cannot understand our perception and the fundamental nature of things absent an understanding of our higher dimensional perspective. The *Particulate Matter* conceptual icon suggests our current understanding of an atom's structure is illusory. It shows a proton on the right consisting of a more densely wound segment of the line of space/time which is a storehouse of potential energy. It also shows an electron cloud on the left consisting of a less densely wound segment of a line of space/time which is a storehouse of kinetic energy. The source of this energy is the unseen displacement of a quantum in the fourth dimension which occurs at a speed equal to that of light. The quantum is manifested as a photon in the expanding three-dimensional subspace. The proton and electron are linearly arrayed on a line of space/time, like words in a sentence or notes on a musical score, the proton appears to be located within the electron cloud when the two are viewed from the fourth-dimensional perspective of our consciousness. This always appears to be the case, irrespective of our position in three dimensions, because the fourth dimensional line of sight is always perpendicular to a plane of being in the three-dimensional subspace. In effect, we create separate, three-dimensional atomic and subatomic particles in the space/time past of our perceptual field because we cannot see their fourth-dimensional connection. This suggests the Standard Model of Particle Physics is the modern equivalent of counting angels on pinheads since, at a fundamental level, particles are merely wound lines of space/time generated by photons which are manifestations of fourth-dimensional quanta in the three-dimensional subspace.

ACI-6: INVISIBLE OFFSET
Allegory No. 6—*Dark Studio*

This conceptual icon is based upon the allegory titled *Dark Studio*. The allegory describes the inner workings of a cineplex consisting of four theaters. Each theater is divided into two studios identified as "A" and "B," respectively. The same movie is shown concurrently in the four theaters. In Studio A the movie runs from the beginning to end, while in Studio B it runs from the end to the beginning. This creates a continuous loop which passes through the projector embedded in the party wall separating the two studios. One can purchase a ticket to Studio A, but Studio B is closed to the public. Therefore, the viewing public is only able to see the film from beginning to end. The Invisible Offset demonstrates metaphorically that for every motion there is an equal and opposite motion. As the first frame of Studio B's future passes through the projector, it simultaneously moves into Studio A's past while the converse is true of Studio B. The white square represents Studio A in which there is light and motion. The black square represents the darkened Studio B. The red loop represents the film in space/

time past which is running clockwise. The blue loop represents the film in space/time future which is running counterclockwise. Each theater represents an Apollonian Universe in which space/time recycles continuously, while the cineplex represents the Apollonian Multiverse.[8]

---

[8] In the interest of simplicity, the conceptual icon does not attempt to illustrate how the different simultaneous showings differ somewhat in content and influence each other in an unending kaleidoscope of probabilities.

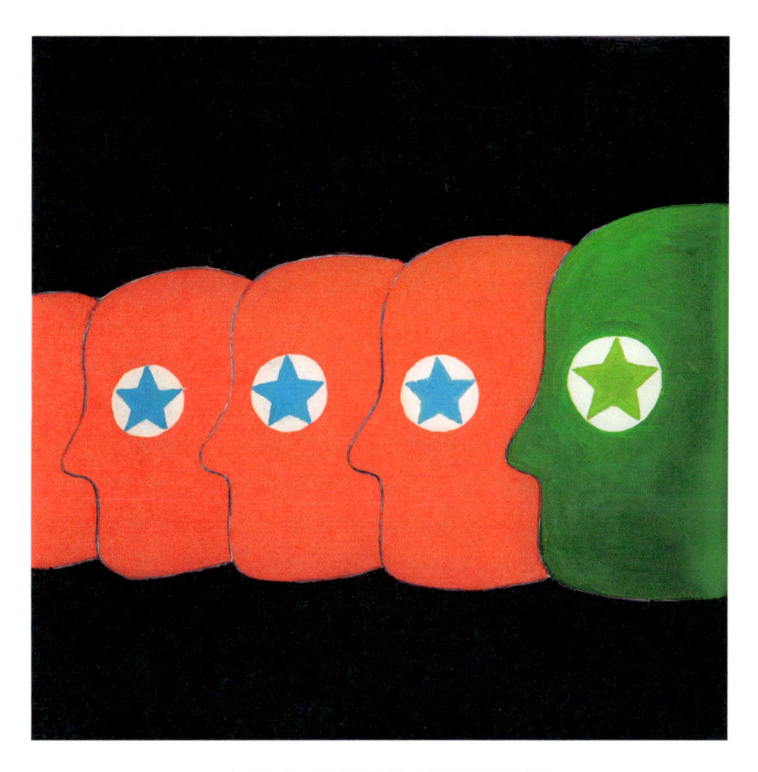

ACI-7: THOU ART THAT
Allegory No. 7—*Infinity Mirror*

This conceptual icon was inspired by the allegory titled *Infinity Mirror*. The allegory illustrates the Apollonian System's principle that individuated conscious can only perceive an external event when the two are at one in being at a fixed point in four-dimensional space. This shared state of being is made possible by prime motion which alternately expands and collapses the perceptual field in space/time separating the observer from the observed. We perceive this flux as the propagation of light. When astronomers look at the light images of distant galaxies captured by the James Webb Space Telescope, they assume they are looking back to a time close to the Big Bang when the universe emerged from nothing as a one-time historical event. Based upon that metaphysical assumption, astronomers have determined the universe's age by means of a regression analysis based upon the universe's observed rate of expansion. This results in an estimated age of approximately fourteen billion years. However, under the Apollonian System's mechanics of perception, they are also affirming the age of their own consciousness since the galaxies were in a shared state of being with the astronomer's consciousness

39

nearly fourteen billion years ago. If the perceptual world is an infinite space/time continuum, our conscious awareness must be eternal and universal. The Ancient Indians clearly understood that our consciousness is not an epiphenomenon of the ephemeral material body as Western science would have us believe. The source of consciousness is a single entity of which our individuated consciousness is a manifestation in the physical world. To this day, Hindus acknowledge this shared divinity by greeting each other reverently with a prayerful gesture. I have named this conceptual icon after a quote from the *Upanishads* which reads "Thou art that." The icon consists of the following: A green head represents the observer's mind-body in the physical world. A green star in the observer's eye indicates his or her consciousness is in a shared state of being with a particular stellar event when perceived. The *Infinity Mirror* allegory illustrates this concept using an old-fashioned barbershop mirror which serves as time's looking glass in which the observer's reflected image is shown as multiple red heads of decreasing size as they recede into the past in accordance with visual perspective. The blue star in each successive image represents an event in transit as it rides a perceptual sphere which is collapsing upon the observer's consciousness. As the light from the star hyper-spherically expands, it moves toward the observer whose consciousness simultaneously hyper-spherically expands in the opposite direction towards the star. This equal and opposite displacement collapses the intervening space/time field separating the observer and observed until they merge in a shared state of being which the mind-body registers on the observer's consciousness as visible light which creates an observable perceptual event.

WORKING CONCEPTUAL ICONS

Description

Working Conceptual Icons demonstrate the dynamics of the Apollonian System's conceptual icons. These dynamics are made possible by the prime motion of an unseen quantum in the fourth dimension at a speed equal to that of light which results in the hyper-spherical expansion of space at each-and-every point in the three-dimensional subspace. We experience this expansion in space/time past of the perceptual world as the propagation of light, passage of time, and rotation.

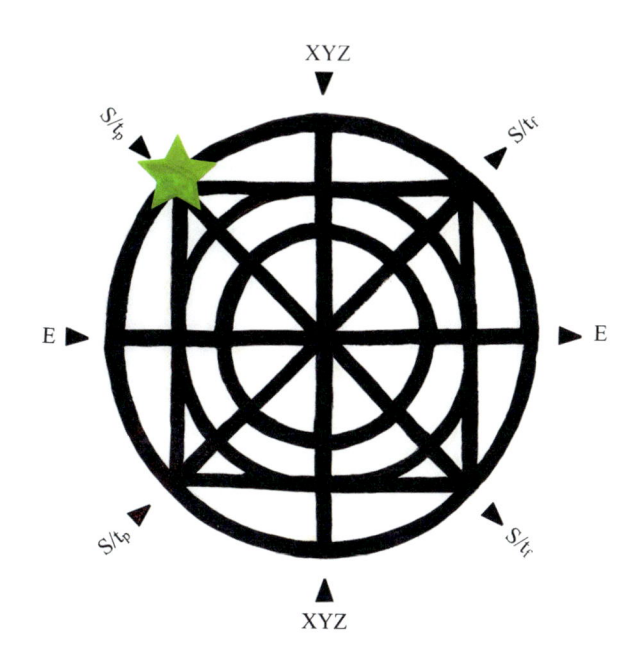

WCI-1: APOLLONIAN COORDINATE SYSTEM
EVENT IN BEING

This conceptual icon displays a color-coded event on the Apollonian Coordinate System. The green star represents an event in a shared state of being with its physical source. The source is defined as a pulsating variable star and the event as the photon flux from a particular pulsation. The star and the photon flux are located at the same point in absolute four-dimensional space relative to an observing consciousness which is similarly fixed at the origin of the coordinate system. However, the observer cannot see the stellar event because the two are set apart in space/time.

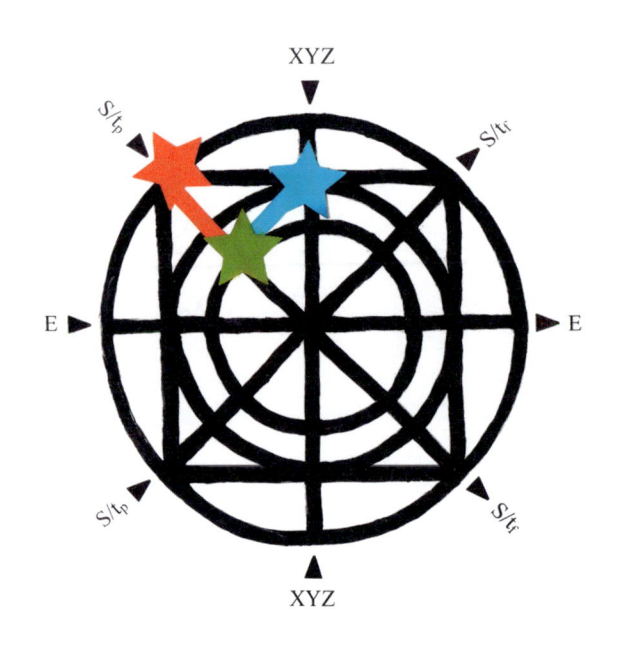

WCI-2: APOLLONIAN COORDINATE SYSTEM
EVENT IN TRANSIT

This conceptual icon demonstrates the paths of a stellar event and its physical source immediately diverge as the source undergoes prime motion in the fourth dimension at a speed equal to that of light. Once again, the source is defined as a pulsating variable star which is represented by a blue star in space/time future, and the stellar event is defined as a flux of light and represented by the green star. The icon depicts an event in transit in which the stellar event leaves behind a red star in space/time past as it rides a perceptual sphere of iso-space/time as it collapses upon the observer's consciousness which is fixed at the origin. As event in transit rides the collapsing sphere, it generates a red line of space/time past which leaves behind a red star marking the event's original location in four-dimensional space. As the blue star representing the physical source moves along its line of existence in the fourth dimension, it generates a blue line of space/time future located perpendicular to the red line of space/time past. This event in transit cannot be seen by the observer since the two are not in a shared state of being.

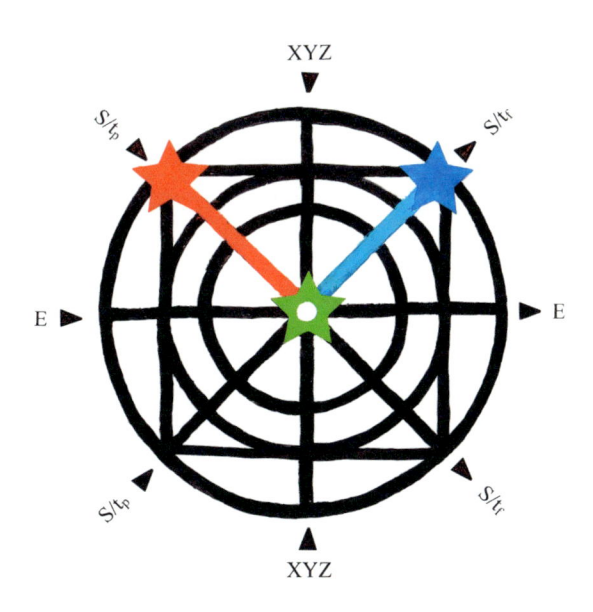

**WCI-3: APOLLONIAN COORDINATE SYSTEM
PERCEPTUAL EVENT**

This conceptual icon demonstrates that a perceptual event, represented by a red star, occurs when the individuated consciousness of the observer, represented by a white dot, and the stellar event, represented by a green star, are in a shared state in being and occupy the same fixed point in four-dimensional space. The green star is linked to the red and blue stars, representing the perceptual event and physical source, by perpendicular red and blue lines of space/time past and future, respectively. Working conceptual icons 1, 2, and 3 demonstrate the mechanics of perception of an event when it is displayed on the four-dimensional Apollonian Coordinate System.

WCI-4: WAVE AND PARTICULATE NATURE OF LIGHT

This conceptual icon accounts for the dual wave and particulate nature of light, the existence of which was established by the "twin split" experiment which demonstrated that a beam of light or subatomic particles exhibits both properties. This characterization is a sophistry in the context of science's historical understanding of the nature of physical reality. However, the paradox is resolved when the experimental results are viewed from the perspective of the Apollonian System in which a mathematical point, known as a quantum, manifests as a photon in the three-dimensional subspace as it undergoes prime motion in the fourth dimension at a speed equal to that of light. We experience this hyper-spherical expansion as the propagation of light and the inferred passage of time, with time being a placeholder for displacement in the fourth dimension which lies beyond our perception. From our single point of view, we perceive this expansion as the rotation of material objects in three-dimensional space or, more accurately, the rotation of three-dimensional space itself as it moves into the fourth dimension as the inertial system we call space/time. While we cannot see the wave form, we can infer its existence under certain experimental conditions and by its periodic manifestation as a photon which occurs every rotation of pi radians of arc. As the rotation collapses the space/time field, it digitizes the wave function which becomes a physical particle at each infinity point. In the *Wave and Particulate Nature of Light* icon, a point in three-dimensional space is shown expanding into the fourth dimension at a speed equal to that of light. While this hyper-spherical expansion is omnidirectional, it is only shown in four directions in the interest of simplicity for ease of illustration. Three intervals of expansion are shown in section as red, green, and blue circles, respectively. The green circle represents the present being of the physical world while the red and blue circles represent iso-space/time past and future, respectively. The white dots embedded in the green circle of the physical world represent the periodic collapse of the wave function which we perceive as a material particle which appears every pi radians of arc of rotation. Since photons and more massive particles are inherently waves, all particles behave in a wave-like manner which accounts for light's dual wave and particulate nature.

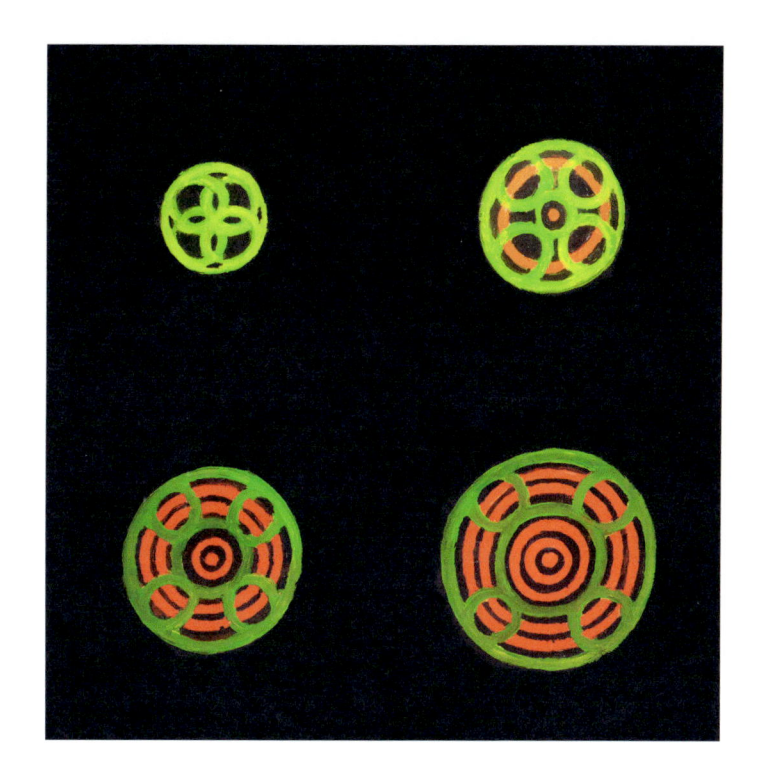

WCI-5: EXPANDING SPHERE OF CONSCIOUSNESS

This conceptual icon demonstrates the following: As individuated consciousness in being (i.e., the self) undergoes prime motion in the fourth dimension at a speed equal to that of light, it expands hyper-spherically until it reaches the four green ring diagrams of equidistant stellar events where it enters a shared states of being and perceptual events occur. This shared state registers upon the observer's consciousness as visible light in accordance with the Apollonian System's mechanics of perception. In the *Expanding Sphere of Consciousness,* the self is shown as a green dot in present being tangent to four green ring diagrams which include four distant stellar events. As the self moves into the fourth dimension, it expands to become a green circle leaving behind a concentric red circle and red dot in relative space/time past while concurrently turning an outlying blue circle in relative space/time future green. As the green circle representing the expanding self reaches the red circle marking the original location of the events, the self enters a shared state of being and a perceptual events occurs. These perceptual mechanics apply concurrently throughout the Apollonian Universe and greater Multiverse.

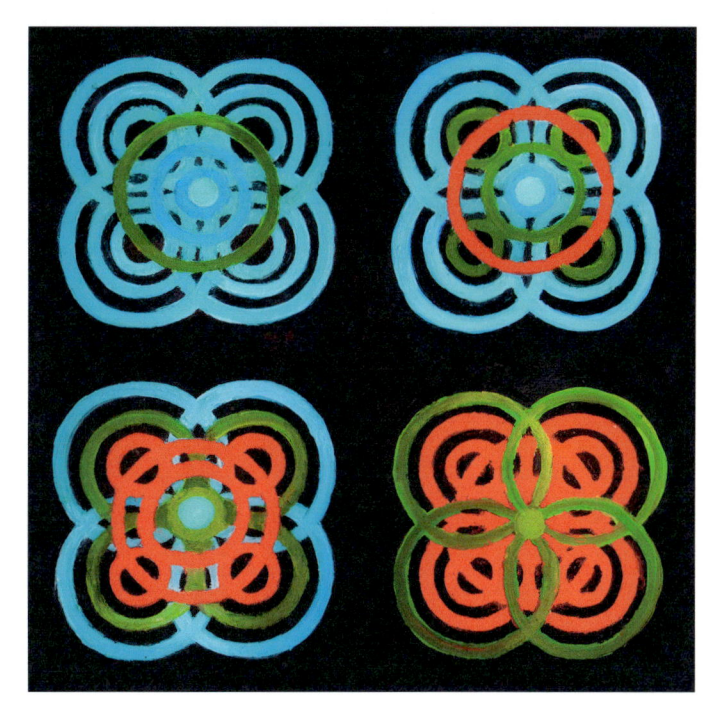

WCI-6: CONTRACTING SPHERE OF PERCEPTION

This conceptual icon places the self at the center of four equidistant stellar events in a shared state of being yet set apart in space/time so they cannot be seen by the self in the present. As the stellar events move into the fourth dimension, their photon flux defines a perceptual sphere which collapses upon the observing self at the speed of light along lines of space/time. When the self and the collapsed perceptual sphere finally enter a shared state of being, the self observes the four perceptual events in visible light. The *Contracting Sphere of Perception* icon shows the initial shared state of being of the observing self and four stellar events as green ring diagrams. As light emitted from the stellar events travels along lines of space/time aligned with the observer, the green sphere of perception collapses upon the observer, leaving behind red, concentric circular sections of spheres of iso-space/time past.

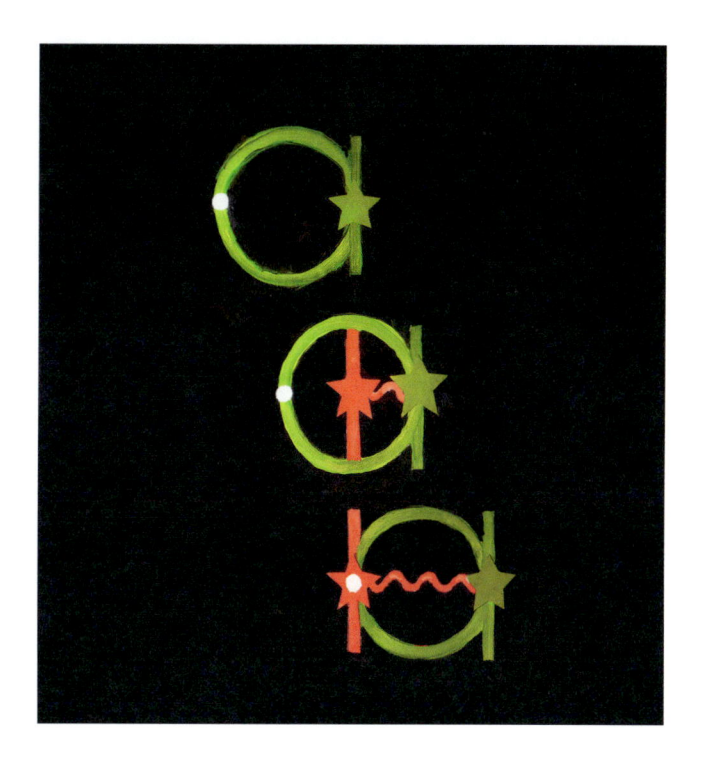

WCI-7: RING DIAGRAM ANALYSIS

This conceptual icon again demonstrates the use of the Apollonian System's ring diagram convention. A ring diagram is created by connecting the ends of a fourth-dimensional line of existence at an infinity point, thereby closing the circle. The first ring on the left includes an observing astronomer's consciousness and a stellar event which form and inertial system undergoing uniform and continuous prime motion to the right at a speed equal to that of light. In the second increment of motion, the observing astronomer has traveled half the distance to the stellar event which is fixed in the plane of being of the physical world where it still cannot be seen. In the third and final increment, the astronomer enters a shared state of being with the stellar event which then becomes visible. Since the astronomer is unaware that he is part of an inertial system, he believes the light from the star is propagating toward him as indicated by the standing sine wave. In the *Ring Diagram Analysis*, the green ring to the left includes the astronomer's consciousness, which is shown as a white dot. The astronomer's consciousness is in a shared state of being with a stellar event, which is shown as a green star representing a single flux from a pulsating variable star. The stellar event is fixed in absolute four-dimensional space and embedded in the plane of being of the physical world shown as a green vertical line. As the ring undergoes prime motion to the right, the plane of being and embedded star turn red when left behind in space/time past. A red, standing sine wave indicates light from the event source has only traversed half the distance to the astronomer. As the ring continues to move to the right under prime motion, the white dot and the red stellar event occupy the same fixed point in absolute four-dimensional space. The astronomer is now able to perceive the event as it existed in the past.

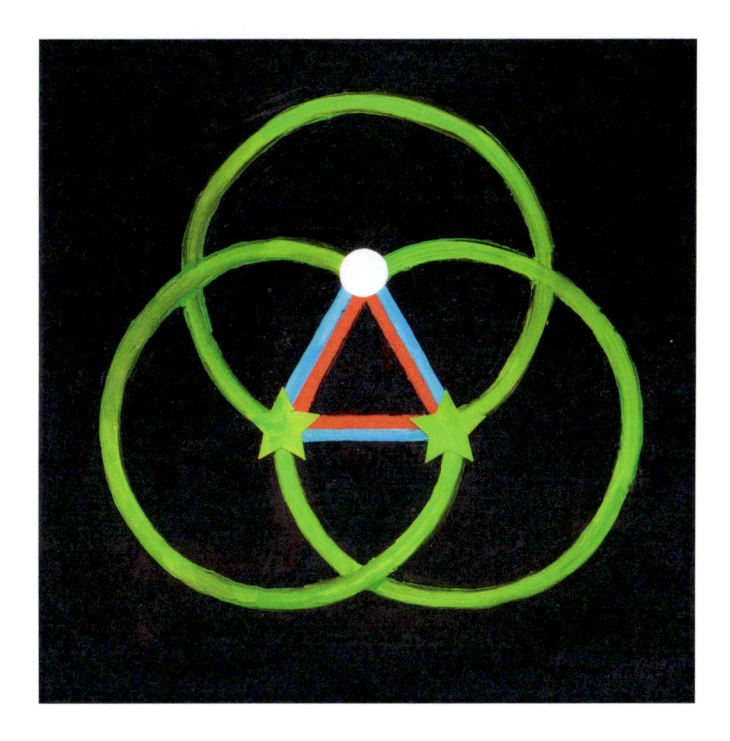

WCI-8: VENN DIAGRAM ANALYSIS

This conceptual icon is in the form of a Venn Diagram consisting of three circular domains in the form of intersecting ring diagrams. In one of the rings, the observing self is shown as a white dot in a shared state of being with two stellar events shown as green stars. In the second ring, one of the green stars is shown in a shared state of being with the white dot and the other green star. Similarly, in the third ring, another green star is shown in a shared state of being with the white dot and the other green star. The white dot and two green stars are connected by red and blue lines of space/time past and future, respectively. In each case, the lines are radii of the hyper-spherically expanding self or hyper-spherically expanding stellar events. These radii are simultaneously self-canceling which collapses the respective perceptual fields, making perception possible.

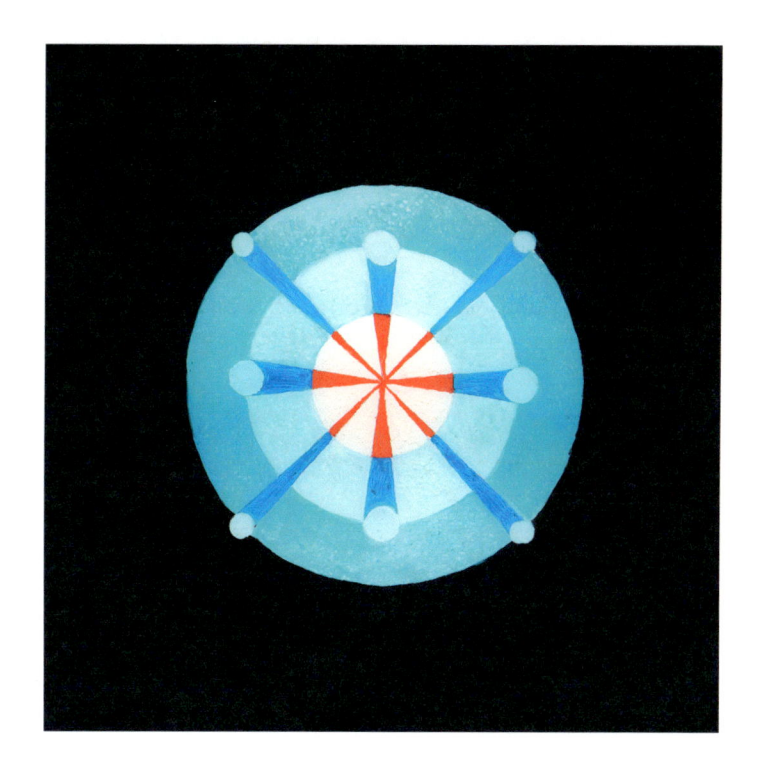

WCI-9: GEOMETRY OF PROBABILITY

This conceptual icon illustrates that probability is a function of four-dimensional geometry subject to prime motion which bombards the self with information in the form of light from the self's own universe and greater multiverse. This flow of information occurs along selected lines of space/time which happen to be aligned with the observing self. The more distant the event, the more converge the lines of space/time become. Consequently, the event contributes an increasingly smaller amount of information used by the observer in the decision-making process. This results in the familiar distribution known as the Gaussian Curve. Each decision made by the observer affects the future universe and multiverse, thereby contributing to their uncertainty. In a sense, the universe is a giant encryption machine in which our actions continually alter the probability of future events making absolute prediction impossible consistent with Heisenberg Uncertainty. In the *Geometry of Probability,* the concentric blue circular areas represent space/time future in which the shade darkens with distance. Similarly, darker blue dots represent more distant events, and lighter blue dots represent closer ones. The central white circular area represents the individuated consciousness of the observer. Light images from the remote events transport information to the observing consciousness along convergent blue lines of space/time future. The observer's consciousness records this input as red sections in space/time past, the areas of which vary as a function of the distance of the sources. The more distant the event, the less impact information about the event has on the observer's decision-making process and the probability of the outcome of future events.

WCI-10: FIBONACCI SPIRAL

This conceptual icon demonstrates that the Fibonacci spiral, often found in the growth patterns of living organisms, is a function of four-dimensional geometry. The icon uses the hyper-spherical expansion of space/time past to account for this natural phenomenon, the spiral pattern of which is explained as follows: When a point in three-dimensional space moves into the fourth dimension, it expands hyper-spherically. The Fibonacci Spiral conceptual icon begins this expansion with a central red dot which becomes a red-orange field representing the space/time past of the perceptual world. Concentric red circles within the red-orange area represent sections of iso-space/time past in which three-dimensional material objects are visible. The radii of the circles represent measures of time, which is defined as the fourth-dimensional displacement of a point in three dimensions which is occurring at a speed equal to that of light. By passing a linear section of the plane of being (assumed but not shown) through the central point, the plane's extension intersects the next outlying circle of iso-space time past. This establishes the point of tangency for the next plane of being. The Fibonacci spiral is generated by repeating this process with respect to each successive circle which produces the spiral form is shown in green as the material form (such as a seashell) is manifested in the present being of the physical world.

WCI-11: SCALE OF BEING

This conceptual icon displays the functional relationship between the will, which is exercised through the agency of the mind-body as thought and physical action, and the consciousness level of individuals, groups, or different species for comparative purposes. The Apollonian System calls the resulting mathematical function the Curve of Aretê which is generated by varying the height of a section of the cone of perception, the volume of which a constant and set by the biophysical limitations of the species. This variance results in the expansion or contraction of the area of the base of the cone of perception. Upon expansion, the area of the base approaches the area of the plane of being as a limit as knowledge is created and accumulated. Consequently, the Curve of Aretê trends upward approaching truth in being asymptotically. The volume of *homo sapiens'* cone of perception is arbitrarily set equal to one which becomes the basis for comparison with other species whose cones of perception have different volumes.

The parameters of the *Scale of Being* include i, j, and k-numbers. The i-number identifies the individual within a species or group within the species. The j-number identifies the species' Curve of Aretê. In the case of Man, the curve is shown as a curved white line. The k-number identifies the level of consciousness. When the volume of a species' cone of perception is less than one, it is incapable of developing culture and civilization. This limitation is described by the diagonal black line in which the will is equal to the level of consciousness which defines the species' natural state of existence. If the k-number is equal to or exceeds one, as represented by a horizonal black line, the species is capable of developing culture which enables it to transcend its bio-physical limitations. This turns the Curve of Aretê increasingly upward as a measure of progress. The red field represents the sum of different species' Curves of Aretê.

CLOSING REMARKS

The Apollonian Exhibition's Folio consists of twenty-two oil paintings of the Apollonian System's conceptual icons. As the artist, I could have achieved greater precision by employing digital technology, but it would have involved the participation of some unknown and long-forgotten programmer. Had Vincent van Gogh lived in the Information Age, he might have used computer graphics as a compositional aid. However, this would have made his painting less authentic, by which I mean they would lack the immediacy, emotional charge, and sense of time and place they now possess with all that implies with respect to their cultural and economic value as works of art. I chose to work in the traditional medium of oil paint because I wanted the Apollonian System's visual vocabulary to be rendered by the hand of its originator with a minimum of intermediation.

From the outset, I imagined the Apollonian Exhibition would become part of a *Triptych Exhibition*. This larger exhibition would incorporate the works of other artists possessing different skill sets but similarly inspired by the Apollonian Exhibition's iconography. As implied by its name, the *Triptych Exhibition* would be in three parts. The first would feature the Apollonian System's two-dimensional paintings of master, allegorical, and working conceptual icons. The second would feature three-dimensional, illuminated glass sculptures I call "light jewelry." (See Exhibit A below for example.) The third would consist of an animated video demonstrating the dynamics of the Apollonian System's conceptual icons. The video's animated images would be inherently four dimensional because motion requires the presence of a fourth dimension of which the viewer is unaware.

Exhibit A

Light Jewel

Example: A glass sculpture of the Oracle of Delphi, also known as the Pythia, sits upon a bronze tripod which rests upon a replica of a limestone block found among the ruins of the Temple of Apollo at Delphi. (The French archaeologists who made the discovery speculated the block's triangular array of notches once secured the feet of the Oracle's tripod.) The frosted surface of the glass obscures the presence of fiber optics cables which illuminate the sculpture from within in a programmed sequence of red, green, blue, and white light. The light show culminates in a scintillating, opalescent effect created by the simultaneous display of color.

    The Oracle holds a bronze ruler in her lefthand symbolizing space and a bronze hourglass in her righthand symbolizing time. In this manner, she embodies the Apollonian System's new paradigm for time, space, and light, with her glass body representing the ephemeral material world and her illumination the eternal light of conscious being. She is flanked on the left by a curved sheet of glass. A play of light upon its etched surface suggests ethylene vapors rising from a glass-filled channel in the limestone block representing a stream of water fed by the sacred Castalian Spring. The Oracle is flanked on the right by a flat sheet of glass. A play of light on its etched surface creates a pattern suggestive of the leaves of the sacred laurel tree. Through art, the ideal finds expression in material form, albeit imperfectly, much as Plato imagined. Art, like engineering, pursues perfection which can never be fully achieved in the physical world, at least from our limited perspective.

APPENDICES

APPENDIX A—Text

Left Panel of the Rear Triptych:

O Delphic oracle,
with eyes white-blank of sun,
tell me of Eternity,
that land of no mirrors,
where all dreamers
with their dreams must vanish
in the Sea of Light.
Tell me also of Aretê,
where truth may be seen
in red, blue, and green,
and mind and body
seek unity in the will
to consciousness.

Sea of Light
by
L. B. Apollonius the Unlearner

Central Panel of the Rear Triptych:

APOLLONIAN SYSTEM

In 1912, the Russian mathematician and philosopher P. D. Ouspensky published his revolutionary book titled Tertium Organum, the Third Canon of Thought in which he states: "I have called this system of higher logic Tertium Organum because for us it the third canon—third instrument—of thought after those of Aristotle and Bacon. The first was Organon, the second, Novum Organum . . ." The Apollonian System is the "third canon of thought" envisioned by Ouspensky but never fully developed by him.

I asked the following question of the night sky while standing on the beach of a remote island in the Western Pacific: *"What is my relationship to the stars above?"* The year was 1953, and I was thirteen years old. Over the next twenty-three years I received the answer in encrypted bits and pieces, with the last piece of the puzzle falling into place in 1976 while I was dining with family and friends on a ranch in Northern California. The last piece was in the form of a comment made by one of the attendees about the nature of perception. In response, I experienced an epiphany I now characterize as seeing the other half of the universe. The mental image consisted of an assembly of colored geometric shapes I later determined established the spatial frame of reference for a dynamic, four-dimensional coordinate system embodying a new paradigm for time, space, and light. I knew instantly I had been given a gift of great value which came with the obligation to understand, document, and communicate to others its informational content. Thus began a forty-plus year metaphysical journey leading to the development of a philosophy I call the "Apollonian System." The Apollonian System uses intuitive mathematics to make the case that consciousness, rather than the physical world, is the ground of being. It concludes we are four-dimensional beings in a conscious multiverse. As such, we enjoy a privileged position in the dimensional hierarchy because the uncertainty inherent in our position makes free will possible. This free will enables us to participate in the creation of our own universe.

In 2022 I published a summary of my philosophical findings in a book titled *Apollonian Coordinate System, the Mechanics of Perception from a Fourth-Dimensional Perspective. Toward a Deductive Metaphysical Science of Conscious Being.* The Apollonian Coordinate System is a mathematical model of Apollonian Universe UA_1, the first in a proposed series of Apollonian Universes UA_n which approach truth in being as a limit. By its existence, UA_1 demonstrates our culture's scientific model of reality is not the only one possible.

While I do not question the value of the scientific method, which Ouspensky called the second canon of thought, I do question the scientific community's interpretation of the experimental evidence. While science appears to be built upon the solid rock of the physical world, it rests upon metaphysical sand for the following reasons:

- Science does not define its terms at a fundamental level. It only describes the functional relationships between undefined terms. (Ouspensky) This is the equivalent of singing in Italian while not understanding the language.

- The physical world cannot be known by perceptual means because our consciousness is the only "thing" we experience directly in a state of being. Therefore, it is the only thing we truly know. Everything else is perceptually derived and subject to interpretation based upon unprovable metaphysical assumptions. Consequently, our a *posteriori* knowledge of the external world is only as good as our a *priori* assumptions about it. (Kant)

- The scientific method's materialist philosophy prohibits science from acknowledging the existence of consciousness in its reality equations. (Bacon) It does so in the name of an objectivity that does not exist because the physical world is subjectively perceived. Absent consciousness, science is "the sound of one hand clapping."

- The inductive reasoning employed by the scientific method cannot prove hypotheses; it can only disprove them. Therefore, all scientific theories remain works in progress until disproven. (Popper)

    I believe science has misinterpreted the experimental evidence because it employs the wrong frame of reference—namely, the space and time of the perceptual world. This makes the current scientific consensus suspect, including such widely accepted theories as the Big Bang, Standard Model of Particle Physics, and Biological Evolution. The Apollonian System changes the frame of reference by assuming the universe is a singularity (i.e., unified field) centered on an observing consciousness fixed in absolute, four-dimensional space. It assumes further the observer's consciousness must be in a shared state of being with an external event for that event to be perceived. That means both must occupy the same fixed point in four-dimensional space. These metaphysical assumptions trigger a kind of counter-Copernican revolution by restoring consciousness to the creative center of the universe in an infinitely centered multiverse consistent with the mathematics of the infinite. Our individuated consciousness is embedded in the three-dimensional physical world, forming an inertial system undergoing prime motion in the fourth dimension at a speed equal to that of light. As such, it establishes the ground of being of the physical world. This means each of us is "in the world but not of it" as Jesus said of Himself.

    I also believe moving the Apollonian System's formal logic to precise mathematics will provide the metaphysics for a new discipline I call "Transcendental Physics"—so-named because it transcends the perceptual world. This Transcendental Physics has the potential of realizing Gottfried Leibniz's unfulfilled dream of creating a *"universal science"* and Nikola Tesla's prediction that *"the day science begins to study non-physical phenomena it will make more progress in one decade than in all the previous time of its existence."* While science and religion cannot be reconciled on their own metaphysical terms, I believe a synthesis is possible at the

Apollonian System's higher-dimensional level.

Apollonian Universe UA_1 rejects the materialism of science's positivist philosophy. It boldly assumes consciousness, rather than the physical world, is the ground of being and the quantum (ψ) is a unit of consciousness (Kant's "thing in itself") rather than a unit of energy as physicists now believe. It does so because our understanding of the external world is inherently subjective as previously stated. For us there can be no objective reality beyond our own consciousness. The quantum resides in fourth dimensional E-space where it hides the light of a photon by which the quantum manifests in the XYZ-subspace. The quantum contains the "DNA" of the physical world which we perceive imperfectly in material form. In that regard, it resembles Plato's ideal. Quantum displacement ($\Delta\psi$) is a result of prime motion in the fourth dimension which occurs at a speed equal to that of light (c). This apparent motion is the source of energy which physicists use to account for change in the phenomenological world. Energy is conserved because the quantum's line of existence in the fourth dimension diverges from the photon's line of space/time in four-dimensional space. This divergence accounts for the constancy of the speed of light, irrespective of relative motion in the perceptual field.

As the quantum undergoes prime motion in the fourth dimension, for every displacement E there are equal displacements X, Y, and Z. This expansion creates the XYZ-space of the physical world, which is tangent to each-and-every point in the manifold of the conscious universe in XYZE-space. The quantum manifests in hyper-spherical space/time as a rotating, spherical section. This wave can only be detected by inference because it is in a state of nonbeing relative to our inertial system. Upon entering a shared state of being with an observing consciousness, the radius of the spherical wave is reduced to zero. This collapses the field to form a physical particle called a photon which conserves the angular momentum of the space/time field as particulate spin. The photon is perceived by the observer's consciousness when the mind-body registers the event's shared state of being which is sensed as visible light within a narrow range of rotational frequencies. The photon's particulate nature is the creation of the mind-body which has a single point of view and is incapable of seeing the fourth-dimensional aspects of events. Nonlocal consciousness, which the Apollonian System calls universal consciousness, looks through the mind-body and down upon the three-dimensional subspace of the physical world from its fourth-dimensional height. The wave's apparent rotation is evidenced by the tangential measure of angular momentum, planetary orbits, swirling about a black hole, oscillating light particles, particulate spin, generation of electromotive force by rotating a conducting loop through a magnetic field, helical pattern of the DNA molecule, and the Fibonacci spiral found in many biological forms, all of which are the three-dimensional expressions of four-dimensional geometry and the mathematics of the infinite which govern the universe and greater multiverse, thereby obviating the need for so-called "natural" laws. At the low end of the frequency spectrum, a photon rotates at a rate of one cycle per second (i.e., one hertz) which has an energy equal to Planck's Constant. The energy level increases in direct proportion to the rate of rotation. The photon is the fundamental building block of the physical world in which each particle's mass, energy, and spectral line profile are a function of the three-dimensional winding density of a segment of a continuous line of space/time in the universe and greater multiverse. The range of the spectrum is set by the speed of light (c), which is constant in our universe and multiverse but differs in parallel universes should they exist, there being no constants in the mathematics of the infinite. In each universe perception and motion are only possible in the space/time past of the perceptual world due to the observer's inability to see the fourth-dimensional aspects of events. The directional nature of prime motion results in a sequence of events we interpret as cause and effect. These perceptual limitations compress space, thereby transforming parallel lines of existence into lines of space/time past which converge upon an infinity point where the direction of prime motion is reversed. This reversal makes the universe eternal in the topological form of a four-dimensional Möbius Strip. The sight lines established by convergent lines of space/time past create the non-Euclidian effect known as visual perspective which prevents us from seeing the present and the other half of the universe in space/time future. The resulting spatial compression digitizes the perceptual world, creating material particles like knots in a quipu string which conserve the potential and kinetic energy of prime motion in the fourth dimension. It also creates the physical experience we call an accelerating force which occurs in gravitational, electric, and magnetic fields. It also masks the simultaneous offsetting of past and future under hyper-spherical expansion, the net effect of which we experience as present being. We find evidence of this offset in the equal and opposite reaction described by Newton in his Third Law of Motion. The Apollonian System characterizes the space/time future of the unseen half of the universe as the world of probabilities, which is the domain of so-called dark energy and matter, the invisible presence of which can only be inferred from their gravitational effects. Since we cannot see the future, it remains uncertain which makes free will possible. Free will enables us to participate in the creation of our own universe, the consequences of which reverberate throughout the multiverse. (Exercising our free will is much like turning a kaleidoscope which creates new patterns.) The multiverse's divergent lines of space/time future determine the probability of outcomes as a function of the choices we make. The inherent uncertainty in our space/time field releases us and others from the bonds of determinism.

When challenging the foundation of a time-tested institution such as science, half measures will not do. Consequently, the Apollonian System becomes a theory of everything by introducing the following:

- New theories of being and knowledge which define all their terms at a fundamental level.
- A dynamic, four-dimensional coordinate system that enables one to track events in four-dimensional space (i.e., "think in eternity") and distinguish perception from reality.
- Conceptual icons which graphically represent the principles of four-dimensional geometry and the mathematics of the infinite for the benefit of those less familiar with mathematical notation or lack an intuitive

sense of the perceptual consequences of imposing their single point of view and three-dimensional perceptual limitations on four-dimensional events.

• A new paradigm for time, space, and light in which time is an analog of unseen motion, perceptual space is dynamic, and the propagation of light is an appearance.

• A philosophical method enabling one to conceptualize alternative Apollonian Universes UAn which asymptotically approach truth in being as a limit.

• A geometric model based upon the mechanics of perception which demonstrates how knowledge is created by means of the dialectic, how cultures and their civilizations evolve in accordance with the mathematical Curve of Aretê by transcending their biophysical limitations, and to what degree different species have realized their consciousness potential on a scale of being.

• The topology of the eternal multiverse as a unified field (i.e., a singularity) in which our consciousness-centered universe is but one of many in which the physical world is finite and the perceptual world is infinite.

• A geometric explanation for probability based upon the divergence of the lines of space/time future in the multiverse.

• The quantum mechanics of light propagation which accounts for the dual wave and particulate nature of light and other seemingly paradoxical effects, including nonlocal communication and the quantum leap's missing time.

• The potential existence of parallel universes made possible by different speeds of light (c_n), there being no constants in the mathematics of infinite space/time.

• A four-dimensional explanation for non-Euclidian visual perspective.

• A four-dimensional explanation for mass/energy equivalence within the continuous atom in which electron shells are storehouses of kinetic energy in the form of loosely wound lines of space/time and the nuclei are storehouses of less accessible potential energy in the form of more densely wound lines of space/time, the sum of the two determining the atom's mass and gravitational effect.

• A hyper-spherical explanation for the growth pattern known as the Fibonacci spiral.

• Metaphysics for a proposed deductive metaphysical science of conscious being which reconciles classical, relativistic, and quantum physics in realization of Gottfried Leibniz's unfulfilled dream of creating a universal science, accomplished by moving the Apollonian System's formal logic to precise mathematics.

• An alternative to the current scientific model of reality in the form of Apollonian Universe UA_1 which is directly governed by a universal consciousness in accordance with the principles of four-dimensional geometry and the mathematics of the infinite, thereby obviating the need for so-called natural laws.

- A theoretical basis for the elusive Unified Field Theory in which the universe is a singularity and particles simply wound segments of lines of space/time past which are perceived as discrete, material particles due to our inability to see the fourth-dimensional connection. In other words, our limited sensory receptivity digitizes the space/time field of the perceptual world.

- Potential reconciliation of science and religion by achieving a synthesis made possible by establishing a higher dimensional context.

APOLLONIAN EXHIBITION

In the closing years of the twentieth century, American surgeon Leonard Schlain wrote a book titled *Art & Physics* which draws attention to the curious relationship between certain art movements and advancements in science. For example, the Cubists attempted to move painting beyond our three-dimensional, single point of view following Einstein's publication of the Special Theory of Relativity which identified space/time as the fourth dimension. While the Cubists succeeded as artists, they failed to achieve their objective, which was to simultaneously view three-dimensional objects from multiple directions—that is to say, from a fourth-dimensional perspective. I attribute their failure to a lack of understanding of the mechanics of perception and their *a priori* assumption the cube is the fundamental building block of space. The Apollonian System assumes the basic unit of space is a four-dimensional sphere. This infinite hypersphere in space/time shrinks to a mathematical point when in a state of being in the physical world into which our consciousness is embedded. Mathematically speaking, this morphology is best described by the equation $0 = \infty$ which means what we perceive as a discrete point in three dimensions is an unbounded hypersphere in four dimensions. We simply can't see its fourth-dimensional aspect. This also explains why some arithmetic solutions involving zero and infinity defy Aristotelian logic, and some numbers are considered "imaginary."

It is not possible to visualize four-dimensional objects because of the limited receptivity of our bodily senses. Consequently, we lack the templates, which Jung call archetypes, to mentally construct their images. However, it is possible to conceptualize them by mathematically projecting our perceptual experience in three-dimensions into a fourth dimension. We do something similar when we conceptualize objects in Euclidian orthographic projection when we can only see them in non-Euclidian visual perspective. This conceptualization is made possible by basing the projection upon the *a priori* metaphysical assumption that the universe is Euclidian, which is the equivalent of lifting ourselves up by our bootstraps. Similarly, the Apollonian Exhibition demonstrates how we can conceive of a reality we cannot and will never see. It does so in twenty-two oil paintings of the Apollonian System's conceptual icons, which are two-dimensions removed from that which they attempt to represent. While not true visualizations, the conceptual icons allow the viewer to think in eternity (i.e., in four-dimensional space), thereby gaining a feeling for four-dimensional objects, much as the fabled blindmen attempting to imagine an elephant, which they cannot see, by feeling its parts. Lewis Mumford said: *"To know a thing by its parts, is science. To feel it as a whole, is art."* The purpose of the Apollonian Exhibition is to create a "feeling" for the whole of Apollonian Universe UA_1. The effort is complicated by the fact that we are part of the universe it attempts to portray. As the Ancient Indians expressed it: *"thou art that,"* I believe a similar notion was responsible for the Ancient Greek admonition *"know thyself"* said to have been inscribed on a column in the Temple of Apollo at Delphi. This admonition is the primary reason I chose the attribution "Apollonian" when naming my philosophy. Only by understanding we are four dimensional, conscious beings in a four-dimensional, conscious multiverse can we understand what we are looking at. This applies equally to the Cubists in the twentieth century and the astronomers in the twenty-first now studying the images sent back by the James Webb Space Telescope.

The Apollonian Exhibition has two parts. The first is labeled "Concept Box." It includes a painting of the Temple of Apollo at Delphi (i.e., the Apollonian), a description of the Apollonian System, a description of this exhibition, an illustrated poem, and copies of my two books, the first of which is titled: *Apollonian Coordinate System, the Mechanics of Perception from a Fourth-Dimensional Perspective. Toward a Deductive Metaphysical Science of Conscious Being.* The second of which is titled: *Apollonian Exhibition, Iconography of the Apollonian System. Conceptualizing the Unseen.* The purpose of the Apollonian Exhibition is to convey a feeling for a more promising reality based in a "reverence for life." It is my hope the Apollonian Exhibition will inspire other artists to translate the Apollonian System's two-dimensional conceptual icons into three-dimensional glass sculptures called "light jewelry" and a four-dimensional animated video (there being no motion in the absence of a fourth dimension). This "Triptych Exhibition" will enable viewers to experience the Apollonian System's alternative reality through different art forms. Sometimes one sees things more clearly from the corner of one's eye.

Lance Burris

Right Panel of the Rear Triptych:

ICONOGRAPHY

MASTER CONCEPTUAL ICONS
    Cosmology
    Ontology
    "Bohr Atom" of the Conscious Universe
    Epistemology

ALLEGORICAL CONCEPTUAL ICONS
    Point of View (Allegory: The Old Man in the Barn)
    Mechanics of Perception (Allegory: Vision at Cobá)
    Creation of Knowledge (Allegory: Nemo's Corollary)
    Evolution of Consciousness (Allegory: Road Race)
    Particulate Matter (Allegory: The Medium is the Message)
    Invisible Offset (Allegory: Dark Studio)
    Thou Art That (Allegory: Infinity Mirror)

WORKING CONCEPTUAL ICONS
    Apollonian Coordinate System, Event in Being
    Apollonian Coordinate System, Event in Transit
    Apollonian Coordinate System, Perceptual Event
    Wave and Particulate Nature of Light
    Expanding Sphere of Consciousness
    Contracting Sphere of Perception
    Ring Diagram Analysis
    Venn Diagram Analysis
    Geometry of Probability
    Fibonacci Spiral
    Scale of Being

COLOR CODE

WHITE: Conscious being in fourth-dimensional E-space. The ground of being of the finite physical world. An admixture of equal intensities of red, blue, and green as the basic components of human color vision.

GREEN: Three-dimensional XYZ-space of the physical world grounded in being by consciousness. The domain of matter and seat of perception where the observer and the observed enter a shared state of being.

RED: Space/time past of the perceptual world in a state of nonbeing. The receding past. The infinite domain of perception and motion.

BLUE: Space/time future of the world of probabilities in a state of nonbeing. The approaching future. The domain of dark matter and energy.

BLACK: Nonexistence.

APPENDIX B-1

The Old Man in the Barn
An Allegory on the Origin and Nature of Individuation

Mr. Wiseman suffered from the cruelties of Alzheimer's disease. Over the years the illness had taken his memory, dignity, and now threatened his very sense of self. Overwhelmed by the old man's increasing dependency, the Wiseman family was forced to place him in an institution located outside the city where, weather permitting, he spent his days wandering aimlessly about the grounds or sunning himself on the south terrace.

Recent testing had revealed Mr. Wiseman's senses were surprisingly acute but registered fleetingly on his consciousness. So advanced was the disease that his retention was reduced to a few seconds. Without hope of cure, the old man seemed destined to spend the rest of his life confined within the walls of the institution, trapped in a kind of timeless bottle, for deprived of his past he could not anticipate the future. Consequently, he lived in the eternal present—an unchanging space of the forever now—devoid of the inference of time. To the old man, life was no longer a moving picture; that is to say, it was no longer a continuous sequence of events linked by cause and effect. Rather, his existence was reduced to an isolated state of being fashioned from clipped stills soon scattered and forgotten. Oblivious to his nightmarish condition, Mr. Wiseman increasingly spent his days seated in a worn wicker chair on the south terrace where, again and again, he smelled the roses for the first time. Such is the fragile nature of the mind and its organic dependency of the self in the perceptual world.

Early one afternoon, events conspired when an attendant forgot to close the south terrace gate. Clad in pajamas, plaid bathrobe, and worn slippers, Mr. Wiseman chanced upon the opening, exited, and shuffled down the path leading to the surrounding farmland. He trudged on with mindless determination until he came upon an abandoned barn. The barn had a curious shape, for it was round with a conical roof topped by a weathervane which veered sharply in the shifting winds of the afternoon. The barn's curved wall was pierced at regular intervals by windows, the panes of which had long been broken out by vandals. Moved by fatigue, the old man entered an open door only to have a sudden gust of wind slam it shut, trapping him within the darkened interior which smelled of hay and rot. Attracted by the light, he instinctively stumbled toward the nearest window like a moth to flame, feeling his way along the wall until he stood transfixed before the opening which framed the image of a solitary oak, bright sky, and scudding clouds. The image glowed softly in his mind's eye like an Impressionist painting hung on a museum wall. After a momentary pause, he moved on. Feeling his way until he stood before the next window where an entirely different picture presented itself. With faded memory, he looked out upon a sparkling new world in which a rusting hulk of a tractor stood in a field of stubble. Again, after pausing briefly, he moved to the next window where he viewed a derelict windmill idling against the sky.

All afternoon, sector by sector, Mr. Wiseman circled the barn. At each window he glimpsed a different reality as the biochemistry of his memory broke down. In this manner, he continued his solitary journey into forgetfulness until later that day he was discovered by attendants from the institution,

Commentary:
How tragic we say of the old man's condition. How tragic to be so alone and uncomprehending that one's relationship to the world is reduced to a momentary view from a window, meaningless in its isolation. Yet are we really that different? Is it not possible we are just as deluded as Mr. Wiseman about the nature of reality? What if we share a common consciousness with others that is linked beyond the physical world? What if the self and other forms of individuation are illusions, the illusory nature of which can be traced to the limitations inherent in the mechanics of perception or, in Mr. Wiseman's case, a chance combination of fading memory, bodily motion, and barn design which create the many from the One in a manner which eludes our detection. Yet, the fact remains, there is only one consciousness in the barn.

APPENDIX B-2

Vision at Cobá
An Allegory on the Mechanics of Perception

In those days Cobá differed greatly from the more popular archaeological sites at Uxmal and Chichén Itzá, for it was an original—a mysteriously private place in its remoteness and semi-abandonment. Upon approaching its ruins, one shares German archaeologist Teobert Maler's sense of discovery when he first stumbled upon the site in the closing years of the nineteenth century. At the time of my visit, less than a half-dozen of Cobá's six thousand structures had been reopened to the tropical sun, and those were but a jumble of limestone blocks chiseled by masons dead for more than a millennium.

It was late afternoon when I set out down the jungle path leading to the Great Pyramid of *Nohoch Mul*. The word "jungle" is inadequate to describe the dense mat of vegetation covering the Yucatán peninsula, the thin soil and underlying limestone of which are incapable of supporting a true rain forest. What the growth lacked in height was more than compensated for by an oppressive density which had a damping effect on the chatter of the many colorful birds and small monkeys populating the area. This was no machete wielding trip, for the path was kept open for tourists. However, one had to be wary of the *cuatro narices*, a local variety of pit viper which occasionally slithered unseen across the shaded path.

With dusk approaching, I was glad to reach my destination where the Great Pyramid thrust skyward through a grasping tangle of vines. Its grand staircase offered the only escape from the claustrophobic vegetation of the plain. The climb up its broken limestone steps was not as difficult as I had imagined, and the view from the top was breathtaking, for a carpet of green extended to the horizon, interrupted only by a constellation of small lakes and a diagonal scar marking the road to Cancún. As my eyes adjusted to the changing light, I saw countless protuberances which I suddenly realized were overgrown structures cast in relief by the setting sun. The thought was staggering. The complex was huge!

As is my habit when visiting ancient sites, I momentarily closed my eyes and allowed my mind to slip backwards, freewheeling in time until it stopped on a day in the tenth century A.D. when Cobá was a functioning part of Maya civilization. I opened my eyes, not as a foreigner removed in time, but as a contemporary of the scene below. The day was bright. The sun, being at high noon, flooded the plain with a harsh light which illuminated the smaller pyramids and temples located in the distance. The stucco sides of the pyramids were painted red, black, white, or yellow ochre, each color indicating a cardinal direction easily referenced by the jungle traveler. In the otherwise crystalline air, a low haze betrayed the existence of thatched huts and burnt cornfields visible at the edge of the scrub forest bordering Lake *Maconxoc*. Dugout canoes skated insect-like on the lake's surface, driven before the wind by outstretched butterfly nets. In the surrounding area, limestone *sacbeob* linked the places of gods and men with elevated pedestrian ways.

Without prompting, I understood why the Maya had built the pyramids at Cobá and elsewhere in Mexico and Central America, for only from the manmade high ground could they escape the suffocating anonymity of the plain. With fire and human labor, they could impose their will upon the resisting environment, constructing villages for shelter, growing maize for food, and building pedestrian ways for access. However, only from atop the pyramid could the Maya see the face they presented to the gods.

Content with these reflections, I began my descent of the Great Pyramid using the zigzagging technique I had developed when descending the heavily restored pyramid at *Chichén Itzá*. The technique was necessitated by the high rise and narrowness of the pyramid's stone steps. Zigzagging provided a more secure footing and a way to avoid the vertigo to which one was susceptible during descent. Thinking of this, I recalled my conversation with the guide at Chichén Itzá, a diminutive Maya woman, barefoot with a small child clinging to her dusty black dress and looking very much as if she had just stepped from the forest. She had that attractive heart-shaped face and bright eyes found among women of her race, and she radiated an intelligence confirmed by her speech, for her words were as carefully arranged as the stones of the reconstructed buildings of which she spoke. While visiting the site's ball court, she ran her small, square hand over a bas-relief depicting a decapitated player. "This," she said, "is the winning captain," her hand touching the kneeling figure, headless and spurting a decorative fountain of blood. "You see," she explained, "it is a mistake to draw a parallel with the gladiators of Rome. My ancestors believed the best must shed their blood for the race to live on. Decapitation was not punishment for failure. It was privilege."

As our conversation continued, I asked if she had heard the pyramids at Tenochtitlán, the Aztec capital in the Valley of Mexico, had been designed to ensure sacrificial victims tossed from the top of the pyramid tumbled freely to the bottom—tactfully omitting, fully tenderized for the chili pot of ritual cannibalism. She responded by saying she had heard that explanation but preferred another. Whereupon she led me back to the base of the pyramid where she pointed to an array of carved, costumed figures. "Here," she said, "the warriors gathered, led by the great lord wearing the long, green feathers of sacred quetzal bird. The lord was accompanied by a bound, sacrificial victim who had been taken prisoner during one of many ritualized wars with the neighboring Maya city states. As the heavily laden procession climbed the pyramid, its members found it necessary to sidestep while ascending the staircase in single file. With each step, the column moved diagonally to the right until it reached the edge of the staircase, whereupon it pivoted ninety degrees to the left. This change in direction was repeated in turn by each follower and reversed upon reaching the opposite edge of the staircase which resulted in an undulating motion up the face of the pyramid. When viewed from below the ascent recreated *Kukulkán,* the feathered serpent god, to whom the pyramid was dedicated. Pure theater, I thought. She had to be right, for as I descended the Great Pyramid of *Nohoch Mul* the ritual sprang from the architecture like music played on an ancient instrument. Upon reaching the bottom of the stairs, I glanced back as if to see the feathered serpent. Instead, I grasped the significance of the structure, for buried within its pyramidal shape was an enduring sign.

Commentary:

And what was that sign? The pyramid is the architectonic expression of a cone and, as such, a model of perception that is activated by ritual. As the warriors advanced up the grand staircase, their serpentine motion formed a standing sine wave along the line of tangency between the face of the pyramid and its underlying cone of perception. At the top of the stairs, the sacrificial stone cradled the apex of an inverted pyramid in the sky. The two pyramids, one seen and the other unseen, represented the past and future, respectively, and the procession's ascent the timeline of the Maya world. The bloody act of sacrifice, with the cutting out of the living heart, discarding and dismemberment of the body, and subsequent cannibalism, symbolized the abandonment of the self and the casting off and recycling of corporality as the old world made way for the new like a snake shedding its skin. By human sacrifice, the priests ushered in a new age, the shedding of blood being the price of admission to the inverted pyramid in the sky which the Maya had to descend for its civilization to live on. And who is to say the priests were not right? Perhaps they stopped the bloodletting in response to declining faith and tribute, thereby closing the door to eternity which brought Maya civilization to its abrupt and mysterious end.

APPENDIX B-3

Nemo's Corollary
An Allegory on the Creation of Knowledge

Nemo D_____ was not your typical high school teacher, for he was a mathematician with a Ph.D. from Berkeley. As such, he was feared by some, loved by many, and respected by all his students in the small valley-town where he had taught mathematics and science for more than a quarter century. "Mr. Dee" as he was called, but never to his face, was of French-Swiss descent. Short in stature, he was surprisingly agile for a man in his sixties, having been a gymnast during his college years. Each morning, he appeared for work dressed in a grey pinstripe suit and woolen ties, the yellow of his plaid tie matching the sheen of his home-pressed coat and tie.

Mr. Dee's face was as wise as it was kindly, for it was deeply furrowed from years of concentration and good humor. His lively hazel eyes were flecked with gold, and his skin was as brown as a nut from the top of his bald pate to the square of his cleanly shaven chin. However, his most distinguishing feature was a sweeping Gallic nose which supported a pair of tortoiseshell half-glasses which could be found pushed back on his broad forehead when not in use. His prominent nose, combined with his pencil-thin, grey moustache and wide, thin-lipped mouth, gave him the appearance of a modern-day René Descartes, which made the moniker "Mr. Dee" even more appropriate.

Mr. Dee knew how to manage the classroom, for he had a hat full of tricks he skillfully deployed on succeeding generations of students, one of which was particularly noteworthy. When not lecturing, he would permit his students to engage in a low level of conversation to dissipate their youthful energy. On occasion the noise would reach and unacceptable level which usually occurred when his back was turned to the class while writing on the blackboard. In response, he would wheel around, seize the yardstick tucked under his left arm and, with a single sweeping motion, bring the flat of the stick down upon the desktop with a resounding "whack." This had the instantaneous effect of bringing the students bolt-upright in their seats. By smoothing his wrinkled brow, his reading glasses would fall from his forehead to the curve of his nose where they glided to a point just short of its cleft tip. Ducking his head, he would survey the glass from side to side, peering intently over the top of the half-glasses. Then, with a twinkle in his eye, he would break into a broad grin. That was sufficient, order having been restored literally by a single stroke, he would resume writing on the blackboard without uttering a word of rebuke.

Like all gifted teachers, Mr. Dee's command of the material and the classroom earned him the respect of the community in which he had become an institution. Mr. Dee didn't just each mathematics and science; rather, he taught the love of mathematics and science. In doing so he was able to dispel the fear of his less able students, making the rigorous subjects more accessible to all. To his students, Mr. Dee was the wise old man on campus and a model of reason. As a consequence, he was consulted on a variety of matters, some of which were of a personal nature, which was the case one spring afternoon when a student approached him after geometry class. The boy of sixteen stood awkwardly in front of his mentor's desk as Mr. Dee vigorously brushed the chalk dust from his brown hands.

"Mr D_____," the boy said with the tentativeness of youth, "may I ask you a personal question?"

The teacher looked up with a bemused smile, "Well, I guess that depends upon the question, doesn't it son? What do you have in mind?"

"It is about religion," the boy said. "I know you and your wife attend church on Sunday, and I am curious…"

Mr. Dee rolled down his shirt sleeves, carefully buttoning each at the wrist as he spoke. "Curious?" he said, his own curiosity having been peaked.

The student continued, "I'm curious how you can be guided by reason and be a religious person at the same time? I mean science and religion don't mix, do they?"

Mr. Dee turned to face the student, "Well now," he said as he struggled into his coat, "that's a serious question and therefore deserving of a serious answer."

The boy was beginning to have second thoughts, sensing he was in for more of a lecture than he had anticipated.

"Let me get this straight. You think science and religion, by which you mean reason and faith, are incompatible. Hmmm. Sounds like you are suggesting that I am being intellectually dishonest when I attend church with my wife."

"No, no," replied the student. "I didn't mean that," he said back-peddling. "It's just you teach mathematics and science which can be proven, while religion cannot be proven and has to be accepted on faith."

"Well, I guess I have a little different take on it, son. You see mathematics, science, and religion all have one thing in common. Do you know what that thing is?"

"No," the boy answered.

"It is uncertainty," Mr. Dee answered.

The boy looked perplexed. "I don't understand."

Mr. Dee continued. "First, I think you need to make a distinction between faith and religion. Faith is personal. I mean you aren't born with faith; you have to find it by yourself. Religion, on the other hand, is institutional. Its purpose is to preserve certain teachings and traditions. You might say religion is like the church building which provides a common root for individual of faith."

Mr. Dee leaned back against the edge of his desk looking intently at the boy. "Now, let me explain what I mean when I say mathematics, science, and religion have uncertainty in common. All are to some extent faith based. Take geometry, for example. Geometry is built upon axioms. Correct? You know what axioms are, don't you son?"

"Yes, they are unprovable assumptions," the boy said, parroting the classroom definition.

"Correct," replied Mr. Dee with obvious pleasure.

"In class you learned you can't do geometry without axioms of some kind. For example, Euclid assumed that parallel lines never intersect. The word "never" implies the lines are infinitely long. Consequently, Euclid couldn't be absolutely certain his assumption was correct because no matter how often he tested the lines, there would always be more to be tested. Therefore, he couldn't prove parallel lines never intersect. He just had to assume it was true until he had evidence to the contrary. Euclid made this and other assumptions to do geometry. While mathematics may be internally consistent, it is always based upon a set of assumptions which cannot be proven or are simply accepted as being self-evident. It isn't necessary these assumptions agree with common sense so long as they are logically consistent. That is why mathematics is, at its heart, uncertain. Science is much the same, although scientists often forget that fact."

"I don't understand. I thought science is based on physical facts," the boy said.

Mr. Dee returned to the blackboard where he retrieved a piece of chalk from the tray. Using his yardstick, he carefully drew an equilateral triangle on the blackboard. "Imagine the blackboard represents all there is, while the triangular area represents all we can know about 'all there is,' because of our perceptual limitations. The area of the triangle is true, but that truth is relative rather than absolute because of the incompleteness of our perceptual information. Do you understand what I am getting at?"

"I think so. You are saying that the blackboard represents the universe, and the triangle represents only that part of the universe we can see," responded the youth.

If by universe you mean the physical world as it exists in three dimensions over time, I would have to disagree. The blackboard represents more than the physical world. It represents all this is, was, and will ever be in n-dimensions. The physical world is simply a slice of a very big and unknowable pie. Some scientists think only things that can be perceived, inferred from perception, and measurable are real. That is a very big assumption since we know our senses are limited. Most think that our consciousness somehow arises out of inert matter, although there is no way to establish that cause and effect relationship any more than it is possible to demonstrate that a radio program originates in the radio."

Mr. Dee, who had just succeeded in knocking down the walls of the boy's world, continued, "As I said, our perceptual limitations restrict the area of the triangle. But that is not our only limitation. Our cultural assumptions determine the triangle's proportions. By cultural assumptions, I mean the axioms used to interpret perception in the process of creating knowledge about the physical world. The axioms of one culture might produce a tall, narrow triangle like this." Mr. Dee drew a second triangle on the blackboard and then continued, "Each of these triangles have the same area representing truth; however, each has different proportions due to their interpretation. Do you understand what I am saying?"

"I think so," the boy replied. "Our body limits what we can see, and our culture shapes how we see it."

"Bingo!" said Mr. Dee. "Our concept of reality is influenced by our cultural frame of reference."

"But how does that apply to religion?" the boy asked.

"Look at it this way, son. Life is a lot like geometry. You can't build a life with assuming something, whether that something can be proven or not. Remember what our friend Archimedes said, "Give me a place to stand, and I will move the Earth." Religious faith gives us a place to stand which helps us deal with the intangibles of life, much as science helps us to deal with the tangibles of the physical world. The greater reality, of which we are a part, must somehow encompass both aspects."

Mr. Dee continues, "You will recall that I said you aren't born with faith; you must find it. But that is not the end of it, for you must reaffirm it each-and- every day based upon your life experience. In other words, faith is no more guaranteed than Euclid's assumption that parallel lines never intersect, or Newton's First Law of Motion always applies. If you find your faith is no longer justified by your experience, you can either abandon it or revisit its underlying assumptions in the hope of reaffirming or restructuring it. Doesn't that sound familiar? Isn't that precisely what we do in mathematics and science? As mortals we have to live with uncertainty and get on with life. My wife and I go to church on Sunday because our faith continues to work for us. It's our skyhook ("skyhook" being one of Mr. Dee's favorite words). Does that answer your question?"

"I guess so," the boy responded, sounding somewhat unconvinced.

"Then I guess my answer will have to do," Mr. Dee said as he wrote "Q.E.D." on the blackboard with a flourish, "because I must be on my way.

As the teacher left the room, the boy reflected on their conversation. While he did not fully grasp all that had been said, Mr. Dee's observations became clearer as the years passed.

Commentary

The allegory highlights the uncertainty inherent in our culturally instilled a priori knowledge used to interpret perception in the creation of our a posteriori knowledge of the physical world. It illustrates that mathematics, science, and religion are faith based to some extent because each is based upon a set of metaphysical assumptions which cannot be proven. Therefore, truth for us can never be absolute and requires constant reaffirmation. To believe otherwise is hubris. The Apollonian System provides a systematic method to establish, revisit, and, if necessary, alter the cultural assumptions underlying our knowledge of the physical world in the context of a greater reality.

APPENDIX B-4

Road Race
An Allegory on the Evolution of Consciousness

The road race took place in the rolling hills outside of Barcelona. Entrants in the 500-kilometer event included Porsche, Lotus, and Ferrari, each manufacturer entering a three-car team. The competition was based upon engineering, maintenance, and driving excellence. The manufacturer, whose team member first crossed the finish line would to be declared the winner.

Within each racing team, the cars had identical chassis design, horsepower, acceleration rate, braking distance, fuel efficiency, and turning radius. However, these characteristics differed from team to team. Each car was equipped with a telephone which permitted the driver to communicate with other members of the team; however, the messages were scrambled to prevent eavesdropping by the competing teams. A unique aspect of the race was the fact that the windshields and side windows of each car were painted black. Consequently, the drivers could only observe the road as it appeared in the rearview mirror.

The race began when the starting light flashed green in the rearview mirror of each car. In response, the drivers advanced ever so slowly, attempting to keep where they were going as close to where they had just been as possible since they were completely blind to the road ahead. The race progressed in this cautious manner for many kilometers until the captain of the Porsche team noticed the following: When his car passed an oak tree on the right, the road tended to the right. When his car passed an oak tree on the left, the road tended to the left. When the car went up or down a hill the road remained straight. These relationships held true until the captain became convinced he had discovered the principles of the road's design. By sharing this observation with his teammates, the Porsche team enjoyed a competitive advantage, and the Porsches moved into the lead, one-by-one disappearing from the rearview mirrors of the competing cars.

As the race progressed, a Lotus driver became aware of the road's design features and communicated this information to his teammates. However, the Lotus team captain directed the drivers to exercise caution in the belief the road characteristics might prove to be a short rather than long wave function since the team lacked sufficient experience to reach any definitive conclusions about its design. This caution proved justified because the rules of the road reversed at 100-kilometer intervals thereafter.

The leading Porsche, which by then was traveling at a very high rate of speed, could not accommodate the sudden change in the road's design and ran off the road and crashed. The remainder of the Porsche team, which was traveling at a somewhat slower speed, was able to make the transition, but doing so with difficulty. The Lotus team was able to adjust to the change because its cars had a shorter turning radius and were traveling at a slower speed. The Ferrari team had no difficulty accommodating the change because it was oblivious to the road design characteristics and was still proceeding cautiously. By the midpoint of the race, one Lotus and one Ferrari car had dropped out of the race for mechanical reasons. When the Lotus team became sufficiently confident about its knowledge of the road's design and greatly increased its speed, it could not overcome the Porsches' commanding lead. Consequently, the Porsche was the first to cross the finish line, the Lotuses second, and the Ferraris third, with Porsche being declared the winning manufacturer.

Commentary:

The road race is a model of biophysical and cultural evolution of different species. It illustrates how consciousness awareness evolves in response to the dialectical process by which perceptually based knowledge and physical reality approach the truth of unity in being. By analogy, racing team members represent the same biological species which are competing intra-specifically, and the different racing teams represent different species competing inter-specifically for survival. The racing car and driver represent the mind-body of each species member. The image in the rearview mirror represents the perceptual world which can only be seen in space/time past, and the road represents the physical world which underlies perception but cannot be observed directly. The scrambling of telephone messages represents the inability of different species to communicate with each other. Conscious awareness of the road's design results in cultural evolution as the increased knowledge is shared with other members of the same species.

The allegory illustrates how species are slow to adapt to change in the absence of cultural evolution. If the change occurs abruptly, this lack of flexibility can prove fatal and even result in the extinction. While technological development can accelerate the rate of cultural evolution, it can also place the species at greater risk because of the uncertainty inherent in our perceptually based knowledge. If the culture's behavior departs from the Curve of Aretê and its spiritual development lags its technological development, the culture becomes incompetent in managing its affairs. This is precisely where we find ourselves today.

Appendix B-5

The Medium is the Message
An Allegory on the Particle as Illusion

Background: I am a painter and writer of books of poetry and philosophy. One of my books contains a poem titled *Atahualpa's Garden*. Atahualpa was the Inca who ruled a sprawling Andean empire at the time of the Spanish conquest of Peru in the sixteenth century and said to have possessed a garden in which the flora and fauna were made of precious metals and stones.

Before writing the poem, I reread William Prescott's classic history, titled *The Conquest of Peru*, which described the garden. While rereading the book I came across the word *quipu*, which means "knot" in Quechua, the language of the Incas. A *quipu* consisted of a bundle of knotted strings used to record and transmit digital information since the Incas had no written language. The color, order, and number of the knots provided useful numeric information. Runners traveled the empire's system of paved pathways wearing *quipus* about the neck while delivering the "mail" via this "pony-less express," there being no horses until the arrival of the Spaniards.

As mentioned, I write books of philosophy. Over the years, I developed a complete metaphysical philosophy called the Apollonian System which provides an alternative to the scientific model of reality. It does so by placing our individual consciousness in the fourth dimension where it looks down upon the three-dimensional physical world. By adopting this elevated perspective, I realized the material world is the illusory product of our perceptual limitations—specifically, our inability to see the fourth-dimensional aspect of things because our consciousness is aligned with that dimension rather than being at least one dimension removed. This lack of receptivity digitizes the space/time continuum of the perceptual world creating the illusion of particles, much like knots on the *quipu* string. By analogy, the string is the medium which consists of a line of space/time, and its undulations are the message perceived as a linear series of knots like words in a sentence. This means particles are simply how we perceive a rotating segment of a line of space/time. As Marshall McLuhan said, "the medium is the message."

Commentary:

The Apollonian System makes the metaphysical assumption that we are four-dimensional, beings, with our consciousness residing in the fourth dimension and our mind-body residing in the three-dimensional subspace of the physical world. Since we must be at least one dimension removed from what we perceive, our single point of view has a fourth-dimensional perspective. Unless we understand the geometry of these mechanics, we have no idea of what we are looking at which has led to the misinterpretation of our perception. Consider the following: Imagine a mathematical point traveling at the speed of light along a line of space/time, Now imagine one segment of the line rotates in the three-dimensional subspace at a greater rate than a later segment. For every displacement of the point in the fourth dimension, there are equal displacements in each of three-dimensional subspace creating hyper-spherical expansion. From our single point of view, this three-dimensional subspace appears to rotate which accounts for the rotation observed in many phenomena, including planetary orbits and particulate spin. As four-dimensional beings, we are unable to perceive the fourth-dimensional aspects of events, so we fill the void with a placeholder we call time, which is the measure of unseen displacement in the fourth dimension.

Now consider the hydrogen atom. The Particulate Matter conceptual icon shows a line of space/time future in which two segments rotate at different frequencies. Let us call the segment with the higher frequency of rotation a proton with the density of rotation determining its mass and gravitation force. Let us call the segment with a lower frequency rate and therefore less mass, an electron. When the two particles are viewed from a fourth-dimensional perspective along the line of space/time, the proton appears to be located inside the orbit of the electron. No matter where we are positioned in three-dimensional space, this always appears to be the case because our point of view and perspective are always perpendicular to the three-dimensional physical world. In this manner, we create the hydrogen atom as it was imagined by Niels Bohr, when in fact, the proton, shown in, and the electron, are linearly arrayed on a line of space/time like knots on a quipu string.

Appendix B-6

Dark Studio

An Allegory on the Unseen Half of the Universe

The Universal Cineplex consists of four theaters built around a central lobby. Each of the theaters is divided into two studios labeled "A" and "B," respectively. A projection machine is embedded in the party wall partitioning each theater. The machine projects the movie onto a screen located in Studio A, wirh the same movie being shown simultaneously in all four theaters. Studio A is a place of light and motion. Darkened Studio B is assumed to be uneventful; however, that is not the case since for every motion of the film in Studio A there is an equal and opposite motion in Studio B. In Studio A the movie is played from the beginning to the end, while in Studio B it is played from the end to the beginning. In the present, the film strip passes through the projector one frame at a time. As the strip passes from Studio B to Studio A, the future becomes the past, and when it passes from Studio A to Studio B the past becomes the future, the vector of time being reversed in both instances. By this process, Studio A is always ready for the next showing. The box office in each theater only sells tickets to Studio A in which the viewer's memory converts a sequence of stills into a continuous narrative from which the viewer infers causality.

Commentary:

The Dark Studio is a metaphor for the dynamics of the Apollonian Multiverse which is in the shape of a four-dimensional Möbius Strip. The surface of the strip represents the infinite perceptual world in which space/time is endlessly recycled. Each theater is a metaphor for the universe as one of an infinite number of universes in the multiverse. The analogy is not complete because only four of an infinite number of universes are considered. All are connected by lines of space/time which transmit information about events in a climate of uncertainty. Studio B is characterized as being dark because the mechanics of perception only allow the observer to see events in the space/time past of Studio A.

Appendix B-7

Infinity Mirror
An Allegory on the Eternal Nature of Consciousness

It was Saturday morning. Walt looked forward to the weekend after a particularly challenging week at the Optical Sciences Laboratory, a precision finishing company located across town from the university from which he had earned his PhD in mechanical engineering. Walt continued to specialize in optical systems after spending months fine tuning the mirrors on the James Webb Space Telescope now in geosynchronous orbit and returning spectacular images of the universe, thanks in part to Walt's handiwork.

It was a blustery, late fall day. The sky was hard blue and the air crisp. Yellow sycamore leaves drifted across the sidewalk as Walt strolled to the barbershop located a few blocks from his brown-shingled house on Campus Street. The "Old-Fashioned Barbershop" was aptly named, as evidenced by its striped barber pole, gold shaving mug logo on the window, and leather barber's chair situated between juxtaposed mirrors. The shop was owned and operated by George Margolis who had barbered most of his life. In later years, he found spending the entire day on his feet increasingly difficult and decided to retire in the expectation that his savings and Social Security benefits would support his modest lifestyle. Upon retiring, he soon discovered he missed conversing with his customers, especially since his wife's passing. He recalled how his father had missed the fellowship he once enjoyed playing backgammon with friends on the Greek island of Tinos. George opened a single-chair barbershop on a reservation-only basis. The shop developed a loyal clientele consisting largely of older men who felt increasingly isolated in the digital age and remembered the time when getting a haircut was a male ritual. George felt there was a niche for a small shop which provided an alternative to today's franchised, unisex beauty salons largely staffed by young women.

Walt entered the shop to the tinkling of an overhead bell. He was warmly greeted by George who was busy sweeping up his last cutomer's hair clippings. Walt soon took his seat on the shop's antique chair.

"The usual?" George asks while draping a protective sheet across Walt's chest.

"The usual," Walt replies, by which he meant a scissor cut and beard trim.

Walter was a regular customer. As such, George knew he had once worked on the James Webb Space Telescope, polishing its mirrors with nano-precision.

"Walt, I thought of you yesterday when I saw an article about the space telescope you worked on. It seems the astronomers are troubled by some of its observations."

"That's right," Walt responded. "Some of its images call the Big Bang Theory into question."

"It doesn't surprise me. The theory never made much sense to me. How can everything come from nothing? I thought it was just a way to avoid the Creator," George being Greek Orthodox.

"In a way you are correct. The scientific method only deals with the material world. It leaves its origin to speculative theories."

"Then the Big Bang Theory isn't science; it is just speculation," George replied with satisfaction.

"That's true of all theories, but the Big Bang provides the best explanation for the existence of the universe given the observational evidence. You might say it provides astronomers with a place to stand as the basis for further study. For example, the Big Bang Theory enabled astronomers to calculate the age of the universe,"

"I understand it is fourteen billion years old."

"Thirteen billion four hundred thousand years, to be more precise. The age was determined by performing a regression analysis using the universe's observed rate of expansion after the Big Bang. When the space telescope observes galaxies more than thirteen billion light years away, they are looking at galaxies which existed at a time close to the Big Bang."

Walt pauses as he looked at his image in shop's infinity mirror in which he sees multiple images of himself of decreasing size as they recede into the past.

"If you think of my image in the mirror as galaxy, you get the idea. When the telescope looks out into space, it looks back in time as well, just like my image in the mirror which is a kind of time machine because it takes time for the opposed mirrors to reflect the images back and forth. Astronomers are seeing galaxies that may no longer exist."

"What if the Big Bang Theory is wrong?"

"Then all bets are off, and astronomers will have to find another explanation."

Garbage in, garbage out, George thought to himself.

"The problem lies in the fact that astronomers interpret their observations based upon the assumption that the Big Bang was as a one-time, historical event; yet they are now observing fully formed galaxies close to the Big Bang when they believe it takes time for gravity to form the stars and the stars to coalesce into galaxies. The required time period extends beyond the Big Bang where nothing should exist."

"What if the theory about star and galaxy formation is wrong?"

"Then the Big Bang Theory remains intact."

"Sounds like turtles standing on the backs of turtles," George replied.

"… with turtles all the way down," Walt responds in acknowledgement of the fable.

"Wouldn't it be a simpler to assume the universe was created by God as religion has long taught?" George asks.

"True, but that is religion and not science. The scientific method's reliance upon perception of the physical world prohibits any consideration of the God hypothesis."

Their discussion having reached its logic end, the two move on to football.

Commentary:

The Apollonian System provides an alternative to the Big Bang Theory. It assumes from our single point of view and fourth dimensional perspective as four-dimensional conscious beings, the universe is best conceptualized as a four-dimensional Möbius Strip in which the space/time of the perceptual world is infinite, and the physical world is finite. This means the so-called Big Bang is merely an inflection point which occurs every rotation of pi-radians of arc are in three dimensions which collapses the perceptual field to an infinity point where the vector of time is reversed creating the illusion of a beginning when space/time is an endless continuum.

Appendix C

EXHIBITION PLAN

The Concept Box is proposed to be mounted on a pedestal with its two triptychs in the open position. Each of the Folio's twenty-two conceptual icons would be illuminated and displayed in the following order:

Master Conceptual Icons
MCI-1
MCI-2
MCI-3
MCI-4

Allegorical Conceptual Icons
ACI-1
ACI-2
ACI-3
ACI-4
ACI-5
ACI-6
ACI-7

Working Conceptual Icons
WCI-1, WCI-2, WCI-3
WCI-4
WCI-5, WCI-6
WCI-7
WCI-8
WCI-9
WCI-10
WCI-11

Appendix D

ABOUT THE AUTHOR/ARTIST

Lance Burris is a fifth generation Californian who lives in Napa, California. He is as a writer, illustrator, and fine arts painter. Lance earned his BA from the University of California at Berkeley and his MS from Boston University. He served as a naval intelligence officer in the Pacific and private computer consultant in Europe before entering the real estate development business in which he served as a project director, city department head, vice president, president, and, finally, president and CEO of his own land development company specializing in large-scale projects.

Lance began writing books of metaphysical philosophy after experiencing an epiphany in the summer of 1976 which he characterizes as "seeing the other half of the universe." Inspired by that vision, he drew upon his studies at Berkeley and his intuition and powers of visualization as an artist to develop a complete metaphysical philosophy he called the Apollonian System, which is based upon the principles of four-dimensional geometry and the mathematics of the infinite. The Apollonian System provides an alternative to the scientific model of reality by assuming that consciousness, rather than the physical world, is the ground of being. It introduces a four-dimensional coordinate system and a new art form which allow one to conceptualize an alternative to our culture's science-based model of reality.

Made in the USA
Columbia, SC
15 November 2024